Applied
Construction
Health and Safety

Theo C Haupt and John J Smallwood

juta

Applied Construction Health and Safety

First published 2023

Juta and Company (Pty) Ltd
First floor, Sunclare Building, 21 Dreyer Street, Claremont 7708
PO Box 14373, Lansdowne 7779, Cape Town, South Africa
www.juta.co.za

© 2023 Juta and Company (Pty) Ltd

978 1 48513 297 4 (Print)
978 1 48513 298 1 (WebPDF)

Project manager: Seshni Kazadi
Editor: Andrew Scholtz
Cover designer: Simplicitas Design
Typesetter: Stronghold Publishing
Indexer: Language Mechanics

Typeset in 10.5 pt on 14 pt Nimbus Roman D

Contents

Abbreviations and Acronyms

ACHASM	Association of Construction Health and Safety Management
ACM	asbestos-containing material
ARCHOSH	Association of Researchers in Construction Safety, Health, and Well-Being
BoQ	bill of quantities
BRA	baseline risk assessment
CBE	Council of the Built Environment
CETA	Construction Education and Training Authority
CF	Compensation Fund
CHSA	construction health and safety agent
CHSMS	construction health and safety method statement
CIB	International Council for Research and Innovation in Building and Construction
CIOB	Chartered Institute of Building
cidb	Construction Industry Development Board
CoA	cost of accidents
COIDA	Compensation for Occupational Injuries and Diseases Act 130 of 1993
CPD	continuing professional development
CPM	construction project manager
CSD	certificate of system design
DEL	Department of Employment and Labour
DPWI	Department of Public Works and Infrastructure
FEM	Federated Employers Mutual Assurance Company (RF) (Pty) Ltd
GBCSA	Green Building Council South Africa
GC	general contractor
H&S	health and safety
HAV	hand-arm vibration
HCA	hazardous chemical agent
HIRA	hazard identification and risk assessment
HSPPUA	Health, Safety, Public Protection and Universal Access
HSM	health and safety manager
ILO	International Labour Organization
ISO	International Organization for Standardization
JSA	job safety analysis
LED	light emitting diode
LO/TO	lockout/tagout
MBA	Master Builders Association
MBSA	Master Builders South Africa

MSDS	material safety data sheet
NHBRC	National Home Builders Registration Council
NIOSH	National Institute for Occupational Safety and Health
ODC	operational design calculations
OCP	operational compliance plan
OH&S	occupational health and safety
OHSA	Occupational Health and Safety Act 85 of 1993, as amended
OHSAS	International Occupational Health and Safety Management Standard
PC	principal contractor
PEPMELF	people; equipment; process/procedure; materials; environment; legal including liability; and financial
PESTLE	political, economic, sociocultural, technological, legal/regulatory and environmental
PPE	personal protective equipment
QCTO	Quality Council for Trades and Occupations
QS	quantity surveyor
SACAP	South African Council for the Architectural Profession
SACPCMP	South African Council for the Project and Construction Management Professions
SAFCEC	South African Forum of Civil Engineering Contractors
Saiosh	South African Institute of Occupational Safety and Health
SAQA	South African Qualifications Authority
SANS	South African National Standards
Scope of Work	Scope of Work for Categories of Registration for the Profession Regulated by the South African Council for the Architectural Profession, 2019
SHERQ	safety, health, environment, risk and quality
SHW	safety, health and well-being
SR	severity rate
SOP	safe operating procedure
SWP	safe working procedure
TCPS	traditional construction procurement system
TQM	total quality management
TVET	technical vocational education and training
UK	United Kingdom
UNESCO	United Nations Educational, Scientific and Cultural Organization
USA	United States of America
WBV	whole-body vibration

Acts, Guidelines and Regulations

Construction (Design and Management) Regulations. UK Statutory Instruments 2015 No 51.

Construction Industry Development Board Act 38 of 2000.

Compensation for Occupational Injuries and Diseases Act 130 of 1993.

Housing Consumers Protection Measures Act 95 of 1998.

Mines Health and Safety Act 29 of 1996.

National Building Regulations and Building Standards Act 103 of 1977.

National Qualification Framework Act 67 of 2008.

Occupational Health and Safety Act 85 of 1993, as amended.

Guidelines and Regulations in terms of the Occupational Health and Safety Act 85 of 1993, as amended:

> Asbestos Regulations, 2001. Government Gazette 23108.
>
> Construction Regulations, 2003. Government Gazette 25207.
>
> Construction Regulations, 2014. Government Gazette 37305.
>
> Driven Machinery Regulations, 1988. Government Notice R: 295.
>
> Driven Machinery Regulations, 2015. Government Gazette 38905.
>
> Facilities Regulations, 2004. Government Notice R: 924.
>
> General Safety Regulations, 2003. Government Notice R: 1010.
>
> Guidelines to the Construction Regulations, 2017. Government Gazette 40883.
>
> Regulations for Hazardous Chemical Agents, 2021. Government Gazette 44348.
>
> South African National Standards (SANS) 10142-1. The wiring of premises. Part 1: Low-voltage installations (Edition 3: 2020).
>
> South African National Standards (SANS) 10400-F. Site operations.
>
> The Scope of Work for Categories of Registration of the Project and Construction Management Professions. Government Gazette 47172.

National Qualifications Framework Act 67 of 2008.

National Building Regulations and Building Standards Act 103 of 1977.

Regulations in terms of the National Building Regulations and Building Standards Act 103 of 1977:

> National Building Regulations, 1990. Government Notice R2378 in Government Gazette 12780.

Project and Construction Management Professions Act 48 of 2000.

Skills Development Act 97 of 1998, as amended.

Workmen's Compensation Act 30 of 1941.

CHAPTER 1

Need for Construction Health and Safety

John J Smallwood

1. Introduction

The need for construction health and safety (H&S) is multifaceted, multi-stakeholder and multilevelled; the three levels being industry, organisation and project. The primary motivators are:

- the prevention of injuries, fatalities and disease
- compliance with legislation and regulations
- religious beliefs and morality
- ethical issues
- humanitarian concerns and respect for people
- the cost of accidents
- synergy, risk and sustainability
- compliance with national and international standards
- corporate social responsibility
- public relations, image and marketing
- adherence with total quality management (TQM) principles
- support of local industry H&S
- the pursuit of better practice.

2. The Prevention of Injuries, Fatalities and Disease

The prevention of injuries, fatalities and disease should be the primary motivator for addressing H&S, which begs the question whether or not an organisation would address H&S if H&S legislation and regulations did not exist? Given that H&S is a project parameter, along with cost, environment, productivity, quality and time, and the negative effect of injuries, fatalities and disease on these parameters, their prevention is critical. For the injured, the family of the deceased and ill workers, injuries, fatalities and disease are a personal issue, and not a statistic, as such. Workers' compensation insurance may compensate to a degree; however, the personal cost in terms of pain and suffering, resultant mental health challenges and the cost to the healthcare system are further issues.

How does the South African construction industry perform? The Construction Industry Development Board (cidb) highlighted the considerable number of accidents, fatalities and other injuries that occurred in the industry in their 2009 report *Construction Health & Safety Status & Recommendations*. The report indicated that the disabling injury incidence rate for the construction industry was 0.98, approximately one disabling injury per hundred, which was 25.6 per cent higher than 0.78, the all-industry average at that time. Furthermore, the fatality rate was reported to be 25.5 per 100 000 workers, which does not compare favourably with the fatality rates of the Australian and United Kingdom (UK) construction industries, namely 3.1 in 2020 and 1.91 in 2020/2021, respectively. However, it should be noted that the UK fatality rate does not include fatalities attributable to motor vehicle accidents (MVAs) in the course of employment and, although it cannot be confirmed, it is unlikely that the Australian fatality rate includes MVAs. The severity rate (SR), in turn, indicates the number of days lost because of accidents for every 1 000 hours worked. The South African construction industry's SR of 1.14 is the fourth highest after the fishing, mining and transport industries, where the all-industry average is 0.59. Given that the average worker works 2 000 hours per year, if the SR is multiplied by 2, the average number of days lost per worker per year can be computed and the construction industry lost 2.28 working days per worker, which equates to 1.0% of working time.

3. Compliance with Legislation and Regulations

Compliance with legislation and regulations constitutes an obvious need for employers to address construction H&S. In terms of the Occupational Health and Safety Act 85 of 1993 (OHSA), in certain instances, contraventions and acts or omissions can result in a fine or imprisonment, or both. During a study *The Holistic Role of Construction H&S* conducted among delegates attending the national H&S conference of a major general contractor in South Africa in 2018, legal compliance by the organisation was ranked second in terms of 34 aspects of the importance of H&S in the context of the organisation and projects.

A further issue relates to enforcement of legislation and regulations during site inspections and the resultant issuing of notices in the case of non-compliance by inspectors of the Department of Employment and Labour (DEL). According to the cidb, inspectors visited 1 415 construction sites during a construction 'blitz' in 2009, which resulted in the issuing of 1 388 (98.1 per cent) notices, of which 86 (6 per cent) were improvement notices, 1 015 (73 per cent) were contravention notices and 287 (21 per cent) were prohibition notices. Furthermore, 52.5 per cent of the contractors were non-compliant. Although compliance does not guarantee an injury-, fatality- and disease-free workplace, it will contribute thereto.

Therefore, H&S legislation may not constitute better practice and numerous examples attest to this fact. First, in terms of section 7 of OHSA, 'Health and safety policy', the chief inspector may direct an employer to prepare a written H&S policy. This is a major shortcoming of the OHSA as the H&S culture of an organisation should be reflected in the H&S policy, which effectively constitutes the foundation for all H&S interventions. Secondly, although section 19 of OHSA, 'Health and safety committees', states that H&S committees should meet as often as may be necessary, but at least once every three months, the latter is inappropriate because monthly meetings should be the minimum requirement, especially in terms of construction projects.

4. Religious Beliefs and Morality

Religion relates to behaviour via morality because behaviour is a function of morality, which, in turn, is a function of belief in and practice of a religion, as asserted by Kate Loewenthal:

$$religion \rightarrow morality \rightarrow behaviour$$

Morality entails subscribing to and practicing ethical standards of behaviour, which includes evaluating intentions and behaviour as right or wrong and good or bad. Moral standards can be rooted in religious tradition and moral laws can be viewed as being divine in terms of origin.

The golden rule establishes a moral level of care for others that we, as humans, are responsible to provide and which is a common theme in most, if not all, of the world's major religions. Basically, it implies that humans should treat other humans as they, themselves, would like to be treated.

Work has been referred to as a deed of spiritual value, which, according to Islam, requires that Allah approve actions and behaviours. The Islamic Tawhidic principles of justice and equity, dignity of labour, and removal of hardship, amplify the need to address H&S.

A related issue is the concept of the economic person, which entails the taking of decisions based on the potential benefits relative to the costs of an intervention, which may result in a decision that conflicts with values and the Tawhidic principles. The Buddhist principle of enlightenment entails the release from picking and choosing, that is, the preference for one thing over another, often at the expense of other people. From a Christian perspective, emphasis on the financial bottom line to measure success can result in unreasonable practices, which can result in hardship and suffering and, consequently, a lack of justice.

Clearly, religions respond to the proverbial comment or question relative to proposed or required H&S interventions, namely: What will it cost? Employers, as with other entities or individuals, cannot put a price on a person's life.

5. Ethical Issues

Ethical business practice includes compliance with legislation and regulations. Furthermore, values embrace ethics and, therefore, H&S legislation and regulations amplify the need for the inclusion of H&S as a value, an organisational performance area and a project parameter. The inclusion of H&S as a value, as opposed to a priority, is important because priorities differ depending on organisational and project challenges, whereas values are entrenched and resistant to change. The failure to include H&S as an organisational performance area and a project parameter is likely to create the impression that H&S is less important than other organisational performance areas, such as financial and the traditional project parameters, namely cost, productivity, quality and time. This, in turn, will marginalise the potential of optimum H&S to be the catalyst for the synergy between the respective project parameters. Therefore, H&S should be afforded status equal to or greater than that afforded to the traditional project parameters of cost, productivity, quality and time.

6. Humanitarian Concerns and Respect for People

Respect for people is one of three principles of *Rethinking construction* initiated by the report of the Construction Task Force chaired by Sir John Egan in the United Kingdom in 1998. The Working Group on Respect for People of the UK's Movement for Innovation (M4I) identified four cross-cutting themes and six action themes. The main overarching cross-cutting theme was the Investors in People Standard, followed by workforce involvement, behavioural issues and an overarching management framework. Committing to the Investors in People Standard is the most effective and systematic means of developing and demonstrating respect for people. Workforce involvement entails the consultation, involvement, engagement and empowerment of all workers, which are prerequisites for workers to be able to contribute to H&S. Although behavioural issues are primarily concerned with the adversarial behaviour in the construction industry, and the impact thereof on performance, the mental health and well-being agenda, which features prominently in current H&S endeavours, amplifies the linkage with H&S's overarching management framework. In terms of an overarching management framework, the business excellence model illustrates the fundamental importance of people as both enablers and results. The enablers highlight the role of people in providing leadership, evolving policy and strategy, and allocating and directing resources, which are all essential within the context of H&S. Results, in turn, include client satisfaction, community acceptance, absenteeism, reportable accidents, injuries, fatalities and disease.

The six action themes of diversity in the workplace; site facilities and the working environment at site level; health; safety; lifelong learning and career development; and the working environment off-site, highlight the linkage between respect for people and H&S.

7. Cost of Accidents

The cost of accidents (CoA) is constituted by both direct and indirect CoA. Although the CoA is a trailing/outcome measure, it is the financial measure that can readily be related to by all stakeholders because it can be expressed as a percentage of project cost or value, organisation business volume or value of construction completed nationally.

Direct costs include those associated with the treatment of the injury, including conveyance to a medical practitioner or a hospital, and any unique compensation offered to workers because of being injured, which is covered by workmen's compensation insurance premiums. The challenge with respect to the provisions of workmen's compensation insurance is that wages are only covered after three days of being off duty or unable to work as a result of the injury.

Indirect costs, which are borne by contractors, include:
- reduced productivity for both the returned worker(s) and the crew or workforce
- clean-up costs
- replacement costs
- stand-by costs
- cost of overtime
- administrative costs
- replacement worker orientation
- costs resulting from delays
- supervision costs
- costs related to rescheduling
- transportation and wages paid while the injured person is idle.

A study conducted in South Africa determined that the direct costs contributed 27 per cent of the CoA and the indirect costs 73 per cent. Another study determined that the total CoA could have been between 4.3 and 5.4 per cent, based on the value of construction work completed in South Africa.

The key issue relative to the CoA is that, ultimately, clients incur the CoA as the CoA is included in contractors' cost structures in the form of indirect costs because contractors do not disaggregate costs when preparing tenders. Therefore, the CoA is a major motivation to project stakeholders to reduce project costs by improving

H&S performance and the resultant synergy between H&S and the other project parameters.

8. Synergy

Synergy refers to the combined power of a group of things when they are working together that is greater than the total power achieved by each working separately. In the context of construction projects, synergy refers to the positive interaction between cost, the environment, H&S, productivity, quality and time. H&S is deemed the catalyst for the synergy because of the following interaction, namely that optimum working platforms engender productivity and shorten production time, which in turn, reduces cost as a result of reduced labour inputs. Such platforms also engender quality because workers have optimum access to the areas in which they are working, which mitigates rework, resulting in increased inputs. Optimum H&S-related interventions mitigate damage to the environment, which mitigates increased costs and time delays.

In terms of synergy, or rather, the negative impact of inadequate H&S on various project parameters, a study conducted on construction project managers in South Africa determined that productivity (87.2 per cent) and quality (80.8 per cent) were the parameters most reported as affecting H&S negatively, followed by cost (72.3 per cent), client perception (68.1 per cent), environment (66 per cent) and schedule (57.4 per cent).

9. Total Quality Management

According to the Associated General Contractors of America, TQM is a continuing process of improvement involving all aspects of an organisation. The main thrust of TQM is continuous improvement in performance relative to H&S, productivity, quality and client and worker satisfaction, and that TQM links the processes related to these project parameters.

10. Risk

Incidents and accidents result in variability relative to the project parameters, as indicated in terms of the negative impact of inadequate H&S, and variability, in turn, increases project risk. The issue of variability is amplified by the outcome of accidents, which is fortuitous as the outcome could be minor, moderate, major or catastrophic. Organisations and projects should actively endeavour to identify, quantify and reduce risk and, although H&S risks are merely one category of construction risk, they should be managed because there is the risk of damage to an organisation's reputation and the tarnishing of its image.

11. Better Practice

Figure 1.1 presents Anglo American plc's H&S journey model. It is notable that compliance, which includes compliance with legislation and regulations, does not constitute the end of the journey but a stage in the journey. Thereafter, the proactive stage precedes the resilient stage, manifested in world class H&S performance – creating a process of continuous improvement/innovation.

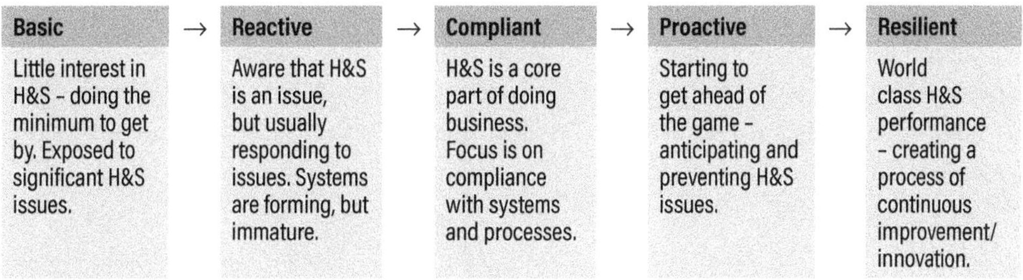

Figure 1.1: Anglo American plc's H&S journey model

12. Sustainability

Accidents can result in fatalities, injuries, disease and damage to material, plant and equipment, all of which result in waste. Waste in solid and other forms impacts on the sustainability of the planet because of the use of landfill sites and the unnecessary consumption of non-renewable resources, in the case of rework. Furthermore, fatalities, injuries and disease threaten the sustainability of the injured or diseased persons' families and their communities. The impact in terms of inadequate H&S also militates against the sustainability of organisations and the industry.

13. Standards

The International Organization for Standardization (ISO) standards, ISO 9001, quality management system, and ISO 14001, environmental management system, both require the implementation of management systems that can complement and support management practices relative to quality and the environment and quality and the environment-related issues being related to H&S issues. The ISO 45000 occupational health and safety management system constitutes the standard in terms of which H&S should be managed. The South African National Standards (SANS) 1393 construction management system standard was developed and introduced to promote and recognise performance improvement by contractors, regardless of size, and is based on recognisable industry standards relative to environmental management, H&S management and quality management.

The contention exists that H&S management and environmental management are aspects of quality management. Given that, in essence, quality means conformance to requirements, such a contention has merit as H&S and environmental legislation and regulations constitute requirements. SANS 1393 constitutes an attempt to integrate the systems, processes and practices relative to the environment, H&S and quality.

14. Corporate Social Responsibility

Standards Australia defines **corporate social responsibility** as a 'mechanism for entities to voluntarily integrate social and environmental concerns into their operations and their interaction with their stakeholders, which are over and above the entity's legal responsibilities'.

The definition suggests that a combination or legislation, regulations and voluntary better practice interventions on the part of organisations is in the best interests of society and all stakeholders. One social concern is H&S, which is logical because H&S-related failures in the form of fatalities, injuries and disease have a wide impact, not just on the individuals concerned.

This, in turn, leads to the concept of triple bottom line reporting, which requires organisations to report on their performance relative to a range of financial, environmental and social indicators, H&S being one of the latter.

15. Public Relations

Public relations is one of nine functions in an organisation and, as a process, can be defined as the deliberate, planned and sustained effort to establish and maintain mutual understanding between an organisation and its public. There are two aspects of public relations, namely internal and external.

The internal aspect of public relations includes the organisation's policies, systems, procedures and protocols; embraces all physical aspects of the organisation, such as the appearance of building(s), vehicles and letterheads, and concerns the personal behaviour of every employee. The internal public relations' need for construction H&S is clear as a result of a range of H&S requirements, courtesy of legislation, regulations and better practice, which include an H&S management system, H&S policy, hazard identification and risk assessment (HIRA), method statements, safe working procedures (SWPs) and H&S protocols such as H&S induction before commencing work on site.

The external aspect of public relations concerns all communication with those stakeholders that are affected by an organisation's activities and upon whose support and goodwill the organisation is dependent. These stakeholders are clients or customers, communities, government, financial institutions, the public, shareholders, subcontractors, manufacturers and suppliers. The external public relations' need for H&S is profound. Better practice clients include H&S as a project value and a project parameter in terms of assessing contractors' performance. Communities are likely to be perturbed by the sustaining of injuries, especially fatalities, and disease involving their members working on sites. The government, in the form of the OH&S Inspectorate of the DEL, is likely to develop a negative image of contractors that are non-compliant in terms of H&S. Financial institutions are likely to view contractors that experience fatalities, injuries and disease as a risk in terms of granting loans, and suppliers similarly in terms of granting credit. The public and shareholders are likely to develop a negative image of contractors that experience fatalities, injuries and disease. Subcontractors are at risk working on a contractor's project that is non-compliant in terms of H&S.

16. Image

Image refers to an impression created at a particular time at a particular level of abstraction. Abstraction, in turn, relates to the amount of complexity in which an entity, object or system is viewed. The higher the level of abstraction, the less detail, while the lower the level of abstraction, the more detail. The highest level of abstraction is the entire system, the next level includes several aspects of the system, while the lowest level includes many aspects of the system. Therefore, a construction organisation, its sites, in the form of the physical construction process, and its activities have the potential to be viewed at differing levels, depending on the stakeholder group and knowledge of the observer. Furthermore, a construction organisation and a site can be viewed indirectly courtesy of the media, while a site can be viewed directly.

Performance relative to H&S affects a client's perception of a contractor's image. Professionalism, which should manifest itself in, among others, an optimum approach to H&S by an organisation, impacts positively on reputation, and a good reputation and external public relations should impact positively on image. Furthermore, external public relations, professionalism, reputation and image contribute individually and collectively to marketing. Clearly, H&S is not solely an operational issue but also a strategic issue because it plays a major role in sustaining the business of construction. A study conducted on award winners that achieved a first, second or third place in the East Cape Master Builders Association's regional H&S competition investigated, among others, the extent to which 44 motivators contributed to their organisations' addressing of H&S. Image was ranked second, professionalism fourth and reputation fifth. These findings, in turn, are underscored by the findings emanating from the study titled *The Holistic Role of Construction H&S*. Organisation image was ranked fourth, organisation reputation fifth, internal public relations eighteenth and external public relations twentieth in terms of the importance of organisation and project H&S.

17. Marketing

Marketing is a further important function in an organisation relative to other functions, such as production, in the form of projects and public relations. An important issue is that of brand, which is a name, term, design, symbol or any other feature that differentiates an organisation's product or service from that of its competitors. Furthermore, a poor image will marginalise an organisation in terms of branding.

A study conducted on the general contractors (GCs) that achieved a place in the Building Industries Federation of South Africa's (BIFSA) national H&S competition and/or received a BIFSA 4- or 5-star H&S grading on one or more of their projects, investigated the marketing benefits of optimum H&S. The study concluded that TQM-related H&S phenomena that contributed to the acquisition of work or additional work clearly indicated the indirect role and benefits of optimum H&S in construction marketing. In essence, optimum H&S provides better practice H&S GCs with a competitive edge and increases their attractiveness to clients.

Findings from the more recent study, *The Holistic Role of Construction H&S,* cited previously, include the fact that 'attractiveness to clients' was ranked third, 'marketing of the organisation' was ranked fourteenth, 'pre-qualification for projects' was ranked ninth and 'attracting clients' was ranked seventeenth in terms of their importance within organisation and project H&S.

One aspect of marketing is that of orientation. A market-oriented organisation places satisfaction at the centre of its philosophy and activities. The organisation will identify how to generate revenue by providing the market with what it requires and then decides how to produce what the market requires at a minimum cost. Within the context of H&S, the issue is that there are clients that view H&S as a value. Furthermore, such clients invariably include H&S as a project parameter and pre-qualify contractors in terms of H&S. Therefore, contractors need to project competence in terms of historical H&S performance and in terms of H&S systems, practices, procedures and protocols. Furthermore, a health and safety conscious contractor is likely to have reduced the cost of construction by attaining the synergy between optimum H&S and the other project parameters, such as productivity, quality, time and the minimisation of the cost of accidents. The production-oriented organisation bases its activities on its products and/or production process, relying on the product or service to sell itself. Despite this approach, optimum H&S is necessary to achieve the required project performance criteria. Historically, contractors have generally been production oriented. However, contractors involved in the housing and property development sectors tend to be more market oriented than materials manufacturers tend to be.

18. Financial Factors and Indicators

Several financial factors and indicators are affected either positively or negatively by H&S performance.

18.1 Profitability

Optimum H&S performance contributes to enhanced profitability because of the containment of the CoA and enhanced the synergy between H&S and the other parameters, which in turn, contribute to a reduction in input costs relative to various resources.

18.2 Share price

A further aspect of profitability is investor confidence, which manifests itself in terms of share prices. The collapse of the incrementally launched Injaka bridge in South Africa in July 1998 resulted in 14 fatalities. Following the news of the collapse, the share price of the contractor undertaking the project fell by 18.3 per cent, from R15.30 to R12.50 a share.

A further notable bridge collapse in South Africa, namely the temporary bridge erected over the M1 motorway in Johannesburg, occurred on a Wednesday afternoon in October 2015 and resulted in two fatalities and 19 injured people. Immediately after the collapse, the share price of the contractor undertaking the project fell by 7.32 per cent, to R11.15, leaving its share price 48.37 per cent lower than it was a year before.

19. Support for Industry H&S Initiatives

Many construction industry organisations, including contractors, materials manufacturers and suppliers and plant and equipment suppliers are members of employer organisations, such as Master Builders South Africa (MBSA), representing master builders associations (MBAs), and the South African Forum of Civil Engineering Contractors (SAFCEC), that provide H&S services and organise H&S competitions relative to their H&S programmes. Membership obligates members to participate and such participation is likely to lead to an improvement in the participants' H&S performance.

20. Review Questions

- Would your organisation address H&S if there was no H&S-related legislation? A philosophical and pragmatic response is required.
- Can the OH&S inspectorate of the DEL inspect H&S in the construction process to the extent that compliance is guaranteed? A philosophical and pragmatic response is required.
- Is compliance with H&S-related legislation sufficient to prevent fatalities, injuries and disease in construction? A philosophical and pragmatic response is required.

21. Review Exercises

- Compute a CoA analysis of three recent construction accidents, preferably with different classes of injury, experienced in your organisation.
- Compile a 750-word article titled *The holistic need for construction H&S*.

Bibliography

Construction Industry Development Board (cidb). 2009. *Construction Health & Safety in South Africa Status & Recommendations*. Pretoria: cidb.

Construction Task Force (Britain). 1998. *Rethinking construction: The report of the construction task force to the deputy prime minister, John Prescott, on the scope for improving the quality and efficiency of UK construction; foreword by Sir John Egan*. London: Department of the Environment, Transport and the Regions.

Eckhardt, RE. 2001. The moral duty to provide for workplace safety. *Professional Safety*, August: 36–38.

Foster, P & Hoult, S. 2013. The safety journey: Using a safety maturity model for safety planning and assurance in the UK coal mining industry. *Minerals*, 3 (1): 59–72.

Grunig, JE. 1993. Image and substance: From symbolic to behavioral relationships. *Public Relations Review*, 19 (2): 121–139.

Health and Safety Commission. 2021. *Workplace fatal injuries in Great Britain, 2021*. London: Health and Safety Executive Books.

Inoue, S. 1997. *Putting Buddhism to work: A new approach to management and business*. Tokyo: Kodansha International Ltd.

Levitt, RE & Samelson, NM. 1993. *Construction safety management*. 2nd edition. New York: John Wiley & Sons, Inc.

Loewenthal, KM. 2000. *The psychology of religion: A short introduction*. Oxford: Oneworld.

Pillay, K & Haupt, TC. 2008. The cost of construction accidents: An exploratory study. In: J Hinze, S Bohner & J Lew (eds) *Proceedings of the CIB W99 14th International Conference on Evolution of and Directions in Construction Safety and Health*, Gainesville, Florida, United States of America (9–11 March 2008): 456–464.

Sadeq, AM & Ahmad, AK (eds). 2004. *Quality management: Islamic Perspectives*. 2nd edition. Dhaka: Islamic Foundation Bangladesh.

Safe Work Australia. 2021. *Key work health and safety statistics Australia 2021*. Canberra: Safe Work Australia.

Slabbert, A. 2015. Murray & Roberts to face the music after bridge collapse. *Moneyweb* 14 October 2015, https://www.moneyweb.co.za/news/murray-roberts-faces-scrutiny-after-bridge-collapse/.

Smallwood, JJ. 2004. Optimum cost: The role of health and safety (H&S). In: JJP Verster (ed) *Proceedings International Cost Engineering Council 4th World Congress*, Cape Town, South Africa (17–21 April 2004).

Smallwood, JJ. 2005. The role of optimum health and safety (H&S) in construction marketing. In: F. Khosrowshahi (ed) *Proceedings of 21st Annual Conference of Association of Researchers in Construction Management (ARCOM)*, SOAS, London, United Kingdom (7–9 September 2005).

Smallwood, JJ. 2014. The motivators for addressing construction health and safety (H&S): A hierarchical perspective. In: R Aulin & A Ek (eds) *Proceedings Achieving Sustainable Construction Health and Safety CIB W99 Conference*, Lund, Sweden (1–3 June 2014): 307–318.

Smallwood, JJ. 2018. The role of construction health and safety (H&S) in the management of the business of construction. In: TA Saurin, DB Costa, M Behm & F Emuze (eds) *Proceedings of the Joint CIB W099 and TG59 Conference 'Coping with the Complexity of Safety, Health, and Wellbeing in Construction'*, Salvador, Brazil (1–3 August 2018): 158–166.

Standards Australia. 2003. AS 8003-2003 Corporate social responsibility. Sydney, Australia: Standards Australia.

Temkin, B. 1998. Concor fundamentals do matter. *Financial Mail*, 10 July 1998: 59.

The Associated General Contractors of America (AGC). 1992. *An Introduction to total quality management*. Washington: Associated General Contractors of America.

CHAPTER 2

Construction Client Brief

John J Smallwood

1. Introduction

The Business Roundtable called upon clients in the United States of America (USA) to be concerned about and involved with construction health and safety (H&S) for the humanitarian reasons of preventing fatalities and injuries, and because of the need for increased attention to H&S in the interests of the economic health of the construction industry. It should be noted that the report was originally published in 1982 and reprinted in 1995.

It is notable that the International Labour Organization (ILO) recommends that clients should coordinate or nominate a competent person to coordinate all activities relating to H&S on their construction projects; inform all contractors on such projects of the special risks to H&S that they are aware of; and require tenderers to make provision for the cost of H&S measures during the construction process.

The Council Directive of 24 June 1992 on the implementation of minimum H&S requirements at temporary or mobile construction sites highlighted the need for and role of clients and designers in construction H&S. The United Kingdom (UK) promulgated the Construction (Design and Management) Regulations, which entrenched client and designer participation in construction H&S in the same way as legislation in other European countries did.

The South African Construction Regulations, 2003 followed the direction set by the UK's Construction (Design and Management) Regulations by including clients and designers in the construction H&S process. The Construction Regulations, 2014 reinforced the 2003 regulations by, among others, requiring clients to conduct a baseline risk assessment (BRA) and basing the H&S specification thereon.

Within the context of construction H&S, the purpose of the client brief is two-fold. First, to inform the client of their duties in terms of the Construction Regulations, and second to determine the client's H&S requirements, if any.

2. Appointment of a Principal Agent, Designers and a Quantity Surveyor

The South African Council for the Project and Construction Management Professions (SACPCMP) recognises six stages of standard services, namely:

Stage 1: Project initiation and briefing

Stage 2: Concept and feasibility

Stage 3: Design development

Stage 4: Tender documentation and procurement

Stage 5: Construction documentation and management

Stage 6: Project closeout

During Stage 1, project initiation and briefing, the design team should be appointed. At the inception of this stage, the principal agent, ideally a construction project manager (CPM) or, failing that, an appropriate agent should be appointed. The irony is that at this stage an uninformed client is likely to be unaware of the Construction Regulations and the designer construction H&S-related requirements. This is exacerbated by non-identification of CPM requirements in the Construction Regulations. Furthermore, if a designer is appointed as the principal agent, it should be noted that the designer construction H&S-related interventions in terms of designers' scope of work in the respective identities of work is not similar to that for CPMs. Furthermore, the designer may not be aware of and/or familiar with the CPM construction H&S-related interventions in the scope of work in the CPM's identity of work.

The principal agent and design team members/practices should be pre-qualified in terms of construction H&S. Pre-qualification and appointment should address the following relative to the applicable stakeholders, namely:

- knowledge of construction practice relative to the project concerned
- ability to work with and coordinate activities of different designers
- construction H&S competencies
- resources
- management systems to be used to monitor the correct allocation of people and resources
- qualifications and experience of staff who will carry out the design duties and functions, both internally and externally resourced
- H&S policy
- H&S management system
- project managing construction H&S practices
- ability to facilitate dialogue and communication between designers and contractors
- designing for construction H&S practices

- quantity surveying construction H&S practices
- project H&S lessons learnt with respect to previous projects.

3. Permit to do Construction Work

In terms of Regulation 5(5) of the Construction Regulations, 2014, clients are required to appoint, in writing, a competent person as an agent when a permit to do construction work is required. A permit to do construction work is required if the intended construction work will exceed 365 days, involve more than 3 600 person days or if the tender value is Grade 7, 8, or 9 in terms of the Construction Industry Development Board (cidb) grading. The maximum value of a contract that Grade 7, 8, and 9 contractors are considered capable of performing is R60 million, R200 million, and no limit, respectively. The permit to do construction work was introduced in the Construction Regulations, 2014 to engender compliance with the Construction Regulations on large projects.

4. Appointment of a Construction H&S Agent

Regulation 5(5) of the Construction Regulations, 2014 states that a client must appoint, in writing, a competent person as an agent where a permit to do construction work is required. However, in terms of Regulation 5(6), a client may appoint, in writing, a competent person as an agent where notification of construction work by the contractor is required.

In terms of Construction Regulation 5(7), an agent must be registered with a statutory body and manage the H&S of a construction project. Given the latter requirement, construction health and safety agents (CHSAs) should be appointed at Stage 1, project initiation and briefing. However, the problem with Regulation 5(5) and (6) is that the term 'agent', as opposed to 'construction health and safety agent', is used, which has been conveniently construed by many stakeholders to mean that a CPM or designer can be appointed as the agent, which, in turn, leads to CHSAs being appointed as subagents. This, in turn, invariably results in CHSAs reporting to and being paid by such agents. Construction Regulation 5(7) exacerbates the situation because it merely states that the appointed agent must be registered with a statutory body, as opposed to stating the SACPCMP. The scope of work for CHSAs relative to Stage 1, project initiation and briefing, in terms of the Scope of Work for Categories of Registration for the Profession Regulated by the South African Council for the Architectural Profession, 2019 is a reminder that no Act or Regulation should be read in isolation because of the following services included therein: '1.2 Assist in developing a clear construction project health and safety brief'; '1.4 Conclude the terms of the agreement with the client'; and '1.6 Advise the client on the adequacy of health and safety competency and resources of the other consultants'. The services to be provided by CHSAs relative to stages 2 to 6 clearly indicate that the 'agent' referred to in the Construction Regulations, 2014 is the CHSA.

Assessment during pre-qualification and/or prior to the appointment of CHSAs should address the following:

- professional indemnity insurance
- general H&S policy
- organisation structure
- allocation of H&S duties
- delegation of H&S responsibilities
- academic qualifications, experience and professional registration
- references from previous clients with respect to, among others, BRA preparation and H&S specification preparation
- overseeing of principal contractors' (PCs) management of H&S
- preparation of H&S files
- previous experience in terms of similar types of contracts and projects, duration and value
- details of H&S resource facilities, such as an H&S library and an H&S management system
- audit trail control mechanisms, standard documents and record-keeping systems
- details of training courses/continuing professional development attended
- knowledge of designing for construction H&S
- procedures for coordination of design H&S issues
- procedures for involving design team in construction H&S management
- examples of H&S specifications
- approach to monitoring competence of PCs
- details of any previous involvement in H&S-related litigation
- details of the likely team to be involved in the project.

5. Baseline Risk Assessment

In terms of Regulation 5(1)(a) of the Construction Regulations, 2014, clients are required to prepare a BRA. BRAs are one of three types of risk assessment and are intended to determine the wide range and status of risks associated with a project, resulting in a risk profile for the project. BRAs are not issue-based or continuous risk assessments, they are overarching in nature and include the risks the client faces in terms of the project concept and during the project phases; the risks, including H&S risks, that the client and designers face in and from the project design; and the residual risks that the client and designers have not resolved that influence construction and H&S. Given that many clients are not specialists in terms of H&S, this requirement amplifies the need to appoint a CHSA.

6. H&S Specification

Regulation 5, 'Client', of the Construction Regulations, 2014 schedules a range of H&S specification-related requirements relative to clients:

(1)(b) prepare an H&S specification based on the BRA

(1)(c) provide the designers with the H&S specification

(1)(d) ensure that the designer takes the H&S specification into account during design

(1)(f) include the H&S specification in the tender documents.

In terms of Construction Regulation 6(1), 'Structures', designers of a structure must, among others:

(b) take the H&S specification into consideration

(c) include in a report to the client before tender stage:

- all relevant H&S information about the design that may affect the pricing of the work
- the geotechnical-science aspects, and the loading that the structure is designed to withstand.

Given the likely implications of such a report and the other requirements of designers in terms of Construction Regulation 6(1), 'Structures', the H&S specification should be amended/revised to include, among others, residual risk and design and construction method statements before including the H&S specification in the tender documents in accordance with Construction Regulation 5(1)(f).

Clients should be sensitised to the rationale or intention of the H&S specification, focus and ideal contents, which are dependent on the CHSA, CPM or principal agent. H&S specifications are not intended to be a regurgitation of the Construction Regulations, 2014, they are intended to be a guide in terms of the risks and the client's H&S requirements that the principal contractor (PC) and contractors need to address.

7. Procurement System and Conditions of Contract

The type of construction procurement system adopted has an impact on construction H&S. The traditional construction procurement system (TCPS), manifested in a fragmented design process; the lack of PC participation in the design process; the preparation of tender documentation by a quantity surveyor (QS) or cost engineer; the competitive tendering process subscribed to by PCs who are committed to H&S to varying degrees; the non-inclusion of detailed H&S preliminaries items in the contract documentation; the lack of reference to H&S standards in the tender documentation; and the invariable award of the project to the lowest price contractor, are not complementary to achieving optimum H&S performance. The design-build procurement system mitigates many of the negative characteristics of the TCPS, especially fragmentation of the design process and the lack of PC participation in the design process. Furthermore, given that constructability promotes H&S, constructability reviews courtesy of the PC will promote designing for construction H&S. However, the benefits of this procurement system are reliant on the commitment of PCs to H&S, which can be addressed through pre-qualification in terms of H&S.

South African standard conditions of contract and documentation are invariably vague in terms of reference to H&S, and the client brief should highlight this aspect. Vague references such as: 'The contractor shall comply with the OSHA, regulations thereunder, including the Construction Regulations' should be avoided. Bills of quantities should facilitate financial provision for construction H&S. In essence, a lump sum or provisional sum should be avoided and a detailed schedule of items should be included in an H&S preliminaries section.

8. Project Duration

This is a further issue, and not just in terms of Stage 5, construction documentation and management, but also in terms of all stages because sufficient time should be allowed for the planning, design and procurement processes. Designers should be enabled to set a goal to 'design complete upon commencement of construction' or preferably upon procurement.

Within the context of H&S, the nature, scope, complexity and value of a project should be considered when deciding on the project duration.

9. Principal Contractors

A range of client requirements relative to PCs are included in Regulation 5, 'Client', of the Construction Regulations, 2014:

(1)(g) requires that clients ensure that potential PCs have made provision for the cost of H&S in their tenders. This is a challenging requirement because a client will require insight into the process and method adopted by PCs in terms of estimating the cost. Given that percentages are dependent on a range of variables, a detailed schedule of H&S preliminary items based on an informed H&S specification is best because it will enable a measured review of the provision for the cost of H&S. Furthermore, a review of a PC's H&S plan in tandem with the review of a PC's provision for the cost of H&S will provide further insight.

(1)(h) requires that clients must ensure that the PC to be appointed has the necessary competencies and resources. This is a further challenging requirement of clients, given that there are eleven construction resources and that the competencies referred to are generic. Consequently, the H&S specification should require PCs to address the following in their H&S plan:

- a summary of the generic and H&S competencies of their management supervision and labour
- a project organogram, including persons responsible for H&S and information in terms of the management thereof
- technology, such as drones, that the PC intends to use
- innovation in the form of say table forms
- the PC's approach to managing materials, including H&S aspects related thereto
- key plant and equipment, and intended subcontractors.

This regulation further amplifies the role of a detailed schedule of H&S preliminaries items based upon an informed contractor H&S specification.

(1)(i)　requires clients to take reasonable steps to ensure cooperation between all contractors appointed by the client, is challenging in terms of the implementation thereof. In terms of (1)(i), the guidelines to the Construction Regulations mention PCs and contractors. First, the 'P' in PC means just that, namely 'principal', that is, there should not be more than one PC, even though the Construction Regulations make provision therefor. However, clients may appoint direct contractors for activities such as installing carpeting, fittings and curtain tracks in hotels. If there are multiple PCs, then (1)(i) alludes to the 'construction management for a fee' construction procurement system, in which case the construction manager manages trade contractors or PCs. The client cannot be expected to ensure cooperation because the client would have to manage such contractors on an hourly basis, including communications to them.

In the case of direct contractors contracting to the client on a project that is subject to the TCPS, the direct contractors should report to the PC in terms of H&S because they will be working in areas worked in by contractors appointed by the PC.

(1)(j)　requires clients to ensure that every PC is registered for workers compensation insurance cover and in good standing, which merely requires a letter from the Compensation Fund (CF) or from Federated Employers Mutual Assurance Company (RF) (Pty) Ltd (FEM). However, the letter of good standing merely confirms registration but does not provide information in terms of performance. The claims ratio based on the percentage workers' compensation insurance claims constituent the of workers compensation insurance paid would be more informative.

(1)(k)　the client is required to appoint every PC in writing, which, in general, should be a logical intervention because it is inconceivable that a client would not conclude a contract with a contractor, notwithstanding the requirement.

(1)(l) the client is required to discuss and negotiate with the PC the contents of the PC's H&S plan and, thereafter, approve it. 'Discuss' and 'negotiate' are not self-explanatory. Furthermore, the H&S plan is intended to constitute a response to the contractor H&S specification. This requirement would require an informed client or CHSA and if the H&S specification is of substance then the process will not be challenging, that is, the H&S plan constitutes a response to the contractor H&S specification. It should be noted that, in terms of the CPM scope of services, CPMs are required to monitor the preparation and auditing of the contractor's H&S plan and approval thereof by the H&S consultant.

(1)(m)that the client is required to ensure that a copy of the PC's H&S plan is available is rhetorical, the key issue here is the contents of the plan and the response to an appropriate contractor H&S specification.

(1)(n) clients must take reasonable steps to ensure that each contractor's H&S plan is implemented and maintained, which is a further challenging requirement of a client. However, (1)(o), which requires that clients conduct periodic H&S audits and prepare documentation at agreed intervals but at least once every 30 days, will facilitate the former requirement. Clients should be sensitised to the need for audits to be of substance and not frivolous, as anecdotal evidence indicates.

(1)(p) the client must ensure that a copy of the H&S audit report is provided to the PC within seven days of the completion thereof. This is a further rhetorical requirement; however, it is necessary in order to establish a timeline. Furthermore, it could be argued that seven days constitutes a whole week, which is a lengthy period on a construction project, a period during which much activity is likely to have occurred without undertaking possible necessary remedial action.

(1)(q) the clients is required to stop any contractor from executing an activity that poses a threat to the health and safety of persons, which is not in accordance with the H&S specification and H&S plan. Doing so constitutes drastic action, however, doing so may be due to the 'life threatening' nature of the threat. Furthermore, stopping an 'activity' implies just that, as opposed to stopping the project. Regardless, the client should ensure that the threat is well documented, including visuals, because PCs may institute claims in response to being stopped from executing an activity.

(1)(r) the client must make sufficient H&S information and appropriate resources available to the PC when changes are made to the design or construction work. The nature of H&S information could include residual risks.

Resources is a broad term, however, it may entail additional project finance as a result of the changes, or an extension of time.

(1)(s) the client must ensure that an H&S file is kept and maintained by the PC. The client brief must impress upon the client the appropriate contents thereof, the rationale therefor and the need to address this in both the designer and the contractor H&S specification.

Regulation 5(2) ensures that sufficient H&S information and appropriate additional resources are available to execute the work safely when requiring additional work to be performed because of a design change or an error due to their actions, which is similar to 5(1)(r). However, this sub-regulation constitutes a reminder regarding the importance of 'freezing the design' and avoiding such changes.

Regulation 5(3) ensures that the contractor provides the provincial director with a report that includes the measures that the contractor intends to implement to ensure a healthy and safe construction site, in the case of a fatality or permanent disabling injury. The client should also be made aware of the requirements of section 24 of the Occupational Health and Safety Act 85 of 1993 (OHSA), 'Report to inspector regarding certain incidents', and also that motor vehicle accidents (MVAs) that occur in the course of employment must be reported to the South African Police Service. It would be erudite to brief the client regarding the reporting of incidents to the CF or FEM.

Regulation 5(4) ensures cooperation between all PCs and contractors where more than one PC is appointed, which is similar in terms of implications to regulation 5(1)(i).

10. Partnering

Partnering is a process that entails establishing working relationships between project stakeholders through a mutually developed formal strategy of commitment and communication with the aim of realising a successful project. Project stakeholders should have a positive disposition towards partnering prior to commencing a project and the process should be initiated at the inception of Stage 1. Partnering should focus on optimising performance relative to the range of project parameters. In terms of H&S, ideally an integrated multi-stakeholder project H&S plan should be developed and implemented. This plan should include the contributions of all project stakeholders, namely the client, CHSA, CPM or principal agent, designers, QS, PC and specialist subcontractors. The partnering process should include H&S interventions during the planning and design, procurement and construction processes, among others.

11. Project H&S Objectives and Goals

Ideally, project H&S goals and objectives should evolve from a partnering process; however, if such a process does not materialise, better practise H&S clients may well have such goals and objectives, which, if the case, should be reinforced and supported by the CPM or principal agent and the CHSA, if appointed. If the client is not better practice H&S oriented, then the CPM or principal agent and CHSA should champion H&S.

Objectives are the specific actions and measurable steps that need to be taken to achieve a goal. These goals could include reducing incidents and accidents, and improving H&S practices on projects. There are three categories of objectives, namely key, critical and specific. A key project objective in terms of H&S is a fatality-, injury- and disease-free project, ie zero incidents. Critical objectives could include management commitment to H&S, H&S education and training and worker participation in H&S. Specific objectives could be management attendance of 100 per cent of H&S committee meetings in the case of management commitment; 100 per cent worker hazard identification and risk assessment (HIRA) training in terms of H&S education and training; and worker participation in 100 per cent of H&S inspections.

12. Client Involvement in the Design Process

Client involvement in and influence of construction H&S during design begins during the client brief. The client's desired building, structure or infrastructure, in the form of its nature, scope and complexity, has a profound impact on construction H&S. For example, a client may require a fifty-storey office block to be erected. Given the predominating type of structural frame in the South African construction industry, namely reinforced concrete, the decision regarding the type of structural frame will be taken accordingly. This, in turn, has a profound influence on construction H&S during the physical construction process and its activities.

Completeness of design is an important issue in terms of H&S, as design variations during the construction process can result in out-of-sequence work, which is detrimental to H&S. Furthermore, completeness of design facilitates planning during Stage 4, tender documentation and procurement, and Stage 5, construction documentation and management, which engenders optimum H&S performance. Consequently, the importance of prompt and timely decision making by clients must be impressed upon the client and designers during the client brief.

13. Contractor H&S Management System/Client–Contractor H&S Requirements

Globally, South Africa included, there are many construction clients that can be deemed better practice construction H&S clients. Their historical contributions to construction H&S set benchmarks and are informed with respect to and underscored the requirements of the Construction Regulations, 2014. Therefore, it is necessary during the client brief to capture and process such clients' construction H&S requirements. Such clients are likely to have a contractor H&S management system that incorporates contractor H&S requirements, while other clients may have contractor H&S requirements but no contractor H&S management system.

- Pre-qualification of contractors is a likely element or requirement and could include:
    □ an H&S management system
    □ an H&S policy
    □ management involvement and participation in H&S
    □ management accountability for H&S
    □ worker participation in H&S
    □ management of H&S throughout the supply chain
    □ leading and trailing H&S indicators.

Clients may set H&S expectations for the other primary project stakeholders in the form of an integrated multi-stakeholder project H&S plan, which schedules the required contributions per stakeholder per stage.

- Clients may:
    □ set H&S objectives and goals
    □ require following of contractor H&S guidelines
    □ have a contractor H&S coordinator who liaises with the PC
    □ incentivise H&S in terms of contractor H&S performance
    □ have pre-tender site meeting protocols with respect to H&S
    □ have site handover protocols with respect to H&S
    □ require digitalised site access control
    □ require contractor employee identification
    □ require attendance of client delivered H&S induction
    □ require contractor drug and other substance testing
    □ have a permit to work system in terms of brownfields projects
    □ focus on H&S during project progress meetings
    □ require submission of contractor H&S meeting minutes
    □ monitor H&S throughout all stages of projects
    □ require PC reporting of incidents to them
    □ require submission of leading and trailing H&S indicators

- participate in accident investigations
- maintain contractor H&S statistics
- require multi-stakeholder project H&S closeout reports.

14. Project Progress Meetings

Clients participate in such meetings as the primary project stakeholder. Therefore, a client should be informed during the H&S component of the client brief of the role of such meetings in terms of managing contractor H&S and improving project H&S performance. H&S aspects that should be addressed during such meetings include:

- leading and trailing H&S statistics
- accidents
- the CHSA's audit
- design originated hazards encountered during construction
- H&S- and resources-related implications of changes to design and construction
- the H&S file.

15. The H&S File

In terms of 'Duties of principal contractor (PC) and contractor', Regulation 7(1)(b) of the Construction Regulations, 2014, the opening of a file that includes all documentation required in terms of the OHSA and Regulations is required. Regulation 7(e) requires the PC to hand over the file to the client upon completion and this file must include a record of all drawings, designs and materials used and other similar information. Regulation 7(f) requires the PC to include a list of contractors appointed by the PC, the agreements between the PC and the contractor and the type of work being undertaken by the contractors.

Furthermore, in terms of Regulation 5(1)(s), clients must ensure that the H&S file is kept and maintained by the PC. This sub-regulation can be debated because the H&S file requires multi-stakeholder contributions and the responsibility is probably best resided in the client or the CHSA. The rationale for the H&S file is a reminder because it must, ultimately, serve as a reference during the use phase of a building, structure or infrastructure. Therefore, the as-laid positions of services should be jointly confirmed by the PC, the contractor that installed the service and the responsible consulting engineer.

16. Project H&S Closeout Report

The client should be briefed with respect to the importance of the closeout report and advised of the potential approaches in terms of ownership and responsibility for the process leading thereto. Ideally, repeat clients should take ownership of the process and, if such clients have an H&S department, the department should be responsible for compiling the report, which includes the coordination of securing the respective stakeholder contributions.

17. Review Questions

- Was the inclusion of client responsibilities in the Construction Regulations, 2014 justified? A philosophical and pragmatic response is required.
- To what extent does the H&S inspectorate of the DEL interrogate client compliance with the client-related requirements of the Construction Regulations, 2014? A pragmatic response is required.
- What aspects of client contributions to construction H&S can be improved?

18. Review Exercises

- Compile a client contractor H&S management programme.
- Compile a 750-word article the titled *The holistic role of clients in construction H&S*.

Bibliography

Commission of the European Communities. 1993. *Four guides for the 'temporary or mobile construction sites' directive*. Brussels: Office for Official Publications of the European Communities.

International Labour Organization (ILO). 1992. *Safety and health in construction*. Geneva: ILO.

The Business Roundtable. 1995. *Improving construction safety performance*. Report A-3. New York: The Business Roundtable.

CHAPTER 3

Construction Health and Safety Policy

Theo C Haupt

1. Introduction

It is common for organisations to start any written contract with a mission statement that makes it perfectly clear at the outset what the construction health and safety (H&S) philosophy is that must be applied to all construction projects that an organisation engages in. This mission statement forms part of the H&S policy of the organisation, which could be developed at both organisation and project level.

A **policy** is defined as 'a set of ideas or a plan of what to do in particular situations that has been agreed to officially by a group of people or a business organisation'. This definition should be kept in mind when forming a construction H&S policy as many organisations manage to go off on a tangent when constructing a policy, either omitting elements that should be included or including elements that are not specifically relevant to the policy itself, or that perhaps belong elsewhere.

There are several reasons for having a construction H&S policy, including:

- ethical or moral reasons, implying that it is the right thing to do because of the moral obligation to provide a working environment that is the safest and healthiest possible
- legal and regulatory reasons, such as captured in the Constitution of South Africa, which require construction organisations to provide and maintain a safe and healthy working environment for all workers
- financial and economic reasons, because of the costs directly and indirectly involved in accidents, injuries and fatalities as a result of construction activities, including insurance premiums, litigation, disability claims, loss of productivity and fines
- practical reasons, which compel construction organisations to commit in writing to construction H&S improvement and goals.

The construction H&S policy is a core document in the management of especially construction H&S. Therefore, it details the structure and framework in which the organisation functions with respect to managing construction H&S. It is also true that having a policy guide that stipulates the 'hows', 'whens' and 'whys' is not enough to achieve excellent construction H&S performance within the organisation and on construction projects. Top or senior management must set the example by not only being committed but also being visibly involved and provide, through their actions, the motivation to comply with and implement the policy and to ensure its continued smooth execution. The role of management extends to the coordinating of various activities so that the organisation works as a cohesive team and contributes, together, to the success of the organisation through the effective management of construction H&S.

A construction H&S policy should be based on the primary aim and vision of the organisation with respect to its H&S loss control goals which include, for example:

- the protection of its workers, general public and all other stakeholders
- zero fatalities
- zero permanent disabilities
- the prevention of all injuries and illnesses due to construction activities
- prevention of any fires, vehicle accidents and property, plant and equipment damage.

The construction H&S policy should make it clear that the intention of the organisation is to make a profit without putting workers or others at any risk of their safety, health and physical well-being. The H&S policy should be public, openly visible and acceptable to all construction workers across the organisation. This policy can be brought to the attention of the workers by, for example:

- displaying it on notice boards on construction sites
- handing out a copy of it to all construction workers
- including it in the induction or orientation programme of new workers
- giving training sessions to existing workers when a new policy is introduced or when existing policy changes
- having a section in the corporate newsletter dedicated to the H&S policy.

Available information, facilities and arrangements in the construction organisation should support the H&S policy. Having these structures in place ensures that the objectives of the policy are achievable. Successful policy implementation depends, to a large degree, on the commitment, involvement and attitude of management towards construction H&S, which can be captured in an H&S policy statement that should be signed and dated by the most senior member of the organisation or project management team.

2. Benefits and Challenges of a Construction H&S Policy

Arguably, a successfully implemented H&S policy will result in high standards of compliance with legislative and regulatory requirements; carefully designed and observed H&S systems of work; construction workers adhering to high standards of cleanliness and housekeeping on construction sites; a safe means of access to all parts of the construction site being kept intact; and toxic, corrosive and flammable substances on the site and in the storing facilities being controlled, where it is relevant or necessary to do so. Furthermore, the H&S policy will contribute to effective consultation about H&S with all stakeholders involved in construction projects and the positive participation by all construction workers. Construction worker H&S training programmes will be well developed and should run on a continuous basis, on all levels, with high degree of participation in them.

The most common challenges with written construction H&S policies are that:
- the policy exists, but few people are aware of it
- the amount of detail required may generate a mass of paperwork
- lack of balance between the policy being simple, straightforward and understandable to all construction workers and omissions, resulting in the failure of the policy
- the policy is simply a restatement of the legal duties and obligations of the construction organisation
- confusion between what an H&S policy is and what an H&S manual is, where the policy in fact refers to the manual on technical issues
- the H&S policy should be shorter rather than longer, as length is not necessarily an indicator of value, appropriateness and usefulness, and it should rather be accompanied by explanatory manuals.

3. Intent of a Construction H&S Policy

In general, the construction H&S policy should demonstrate:
- the necessity for active leadership from top and senior management; direct participation and involvement of all levels of management; and the enthusiastic support of the entire organisation
- the intention to integrate loss control into all operations, including compliance with all applicable construction H&S standards, laws and regulations
- the visible and tangible commitment and involvement of the management of the organisation in construction H&S
- the importance that the organisation places on the H&S of construction workers,
- the emphasis placed by the organisation on efficient construction operations, with a minimum of incidents, accidents, fatalities and losses.

4. Content of a Construction H&S Policy

For the construction H&S policy to be effective, it must include, inter alia, the following elements in some shape and form:

- a general H&S policy statement
- statement of the basic objectives and detailed roles and procedures to cater for specified hazards, and provide a framework for setting construction H&S goals and objectives
- the commitment and involvement of top management to a safe and healthy construction workplace, the alignment of construction H&S principles with the strategic direction of the organisation, planning objectives and measurement of the results
- defined duties and extent of responsibility for construction H&S of specified line management levels, while also clearly identifying who has overall responsibility for construction H&S in the organisation
- a clear definition of the functions of various construction H&S personnel and their relationship to line and senior management
- a commitment to grow stakeholder and worker engagement and commitment through meaningful consultation on critical construction H&S aspects in the organisation and on its construction projects, including the construction H&S policy itself
- a statement confirming the responsibility of all workers and stakeholders to uphold the construction H&S principles of the organisation and the maintenance of a safe and healthy construction workplace
- the identified and prioritised hazards, as well as the precautions to be taken to mitigate them, including the roles of workers, visitors and contractors
- details of how the construction H&S policy will be reviewed and improved, where necessary
- a system for supervising construction H&S performance and how information about such performance will be communicated, including what information system will be used to monitor the implementation of the construction H&S system and identify needs that arise
- a description of how the construction H&S activities will be resourced and funded by the organisation, including budgeting process,
- a statement of how compliance with construction legislation, regulations, codes of practice and standards will be achieved.

5. Structure of a Construction H&S Policy

Typically, the construction H&S policy addresses these elements in three major sections, namely:

1. Statement of intent, including specifying objectives (H&S policy statement):
    - outlines the overall philosophy of the organisation in relation to the management of construction H&S.
2. Organisational details (details of the people and their allocated duties with respect to construction H&S involvement):
    - clearly indicates who is responsible to whom for what in the form of a management structure diagram or organogram because everyone within the organisation with construction H&S responsibilities and duties needs to have clear and specific roles, responsibilities, functions and accountabilities
    - specifies how and when the construction H&S policy implementation is to be monitored, evaluated and reviewed
    - indicates the functions and responsibilities of construction H&S committees and representatives
    - shows how construction H&S and associated accountabilities are addressed when recruiting and employing construction workers.
3. Organisational construction H&S systems and procedures:
    - regarded as the primary part of the construction H&S policy as it spells out at operational level how construction H&S is to be maintained at all levels throughout the organisation
    - includes details of, for example:
        - procedures that cover Covid-19 protocols, such as screening, testing, referrals, distancing and working remotely
        - construction H&S training, audits, accident reporting and investigation
        - safe systems of work (standard working procedures or standard operating procedures), permit to work systems, noise and environment control
        - utilisation of personal protective equipment (PPE), fire H&S
        - liaison with contractors, manufacturers and suppliers
        - machine guarding, emergency procedures and medical
        - worker welfare considerations.

6. Drafting a Construction H&S Policy

It is important to be aware that a construction H&S policy becomes a legally binding document for everyone in the organisation, from top or senior management to construction workers. Unfortunately, construction H&S policy statements have historically been lengthy and long-winded, although a one- or two-page policy document would unlikely contain the detail that is required, regardless of the size and complexity of the organisation or construction project.

Note that:

- The Occupational Health and Safety Act, as amended, does not require employers to have an occupational health and safety (OH&S) or construction H&S policy, unless directed to do so by the Inspectorate of the Department of Employment and Labour. This generally happens if the employer is working in a particularly high risk industry.
- The Mines Health and Safety Act 29 of 1996 does require every mine to have an OH&S policy.

The policy requirements of, for example, the International Occupational Health and Safety Management Standard (OHSAS) 18001:2007, being replaced with International Organization for Standardization (ISO) 45001:2018 and ISO 14001:2017, provides excellent guidance on the formulation and content of construction H&S policies.

6.1 H&S policy statement

A word of caution on the H&S policy statement is necessary. Since the organisation will be held accountable for the commitments that are made in the policy statement, it should not make any commitments that it has no intention of adhering to or that it cannot live up to. Therefore, top management should be involved in defining the policy statement and ensure that the construction H&S policy statement is aligned to the construction H&S objectives and targets of the organisation. In other words, the organisation may not have a policy statement without corresponding objectives and targets, nor may objectives and targets be formulated that are not referenced in the policy statement. It is critical to note that the policy statement may only be generated once all the hazards that are typically encountered in the daily operations of the organisation have been identified and assessed.

The construction H&S policy statement should preferably fit on a credit card sized business card that can be carried in the shirt pocket of any construction worker. It should be concise, unambiguous, flexible and short enough for any construction worker to be able to recite it in full. Note that evidence of a positive construction H&S culture is seen when every construction worker knows and can repeat the construction H&S policy of the organisation when requested to do so by anyone.

It is suggested that the following be included in a H&S policy statement:

- a clear statement of intent to provide safe and healthy working conditions and environment on all construction projects
- a declaration that construction activities will not negatively impact others or the environment

- arrangements for consultation with and engagement of construction workers
- a commitment to adequately resource construction H&S; to comply with all applicable legal and regulatory requirements; to cooperate with all workers; and to inform all stakeholders and workers about the H&S policy.

An example of construction H&S policy statement is:

> As a company we are committed to construction health and safety as a value, integrate health and safety into all our construction activities so that we do not negatively affect others and the natural environment, consult regularly with our construction workers, provide adequate health and safety resources, comply with the law and keep all our workers and the public informed about health and safety.
>
> John Doe – CEO of ABC Construction

6.2 Minimum requirements of a construction H&S policy

An effective construction H&S policy should not be an off-the-shelf policy but one developed to address and respond to the particular specific needs of the construction organisation. Legal compliance is but one requirement of a successful H&S policy. Standard commercially available policies may fall short of the practical and legal requirements of the organisation and the construction projects it engages in. Both H&S responsibility and accountability must be defined. Accountability refers to responsibility that is evaluated and measured on a regular basis by the responsible party. Some of the inclusions include:

- clearly articulated management commitment to and involvement in the full life of the construction H&S policy and management system:
  - develop, document, implement, maintain and improve the construction H&S policy and management system
  - availability and adequacy of resources
  - allocation of roles, responsibilities and authority of named persons within the overall management structure of the organisation and construction project
- mandatory commitment statements:
  - commitment to reviewing and the continual improvement of construction H&S performance
  - commitment to preventing environmental pollution, ill health, injury and fatalities from construction activities
  - commitment to complying with the prevailing legislative and regulatory framework and other requirements
  - commitment to effective and transparent communication to all stakeholders

- appropriateness and relevance of the H&S policy statement:
  - appropriate to the purpose of the organisation in terms of what the core construction business and activities of the organisation are
  - appropriate to the nature and scale of the construction activities of the organisation.

7. Examples of Construction H&S Policies

The following are examples of construction H&S policies of varying lengths.

Example 3.1: Example of a construction H&S policy

JOHN DOE SECURITY CONSTRUCTION HEALTH AND SAFETY POLICY

Management and Responsibility

We proactively manage construction health and safety as an integral part of our construction and engineering business, operations and practices.

We require our regional and local management structures to fully comply with all the applicable laws, and the internal and external construction health and safety requirements that affect our operations.

Control and Auditing

We only operate processes, apply construction technologies and methods that are assessed for their construction health and safety risks.

All construction materials, plant, tools and equipment are assessed for their regulatory compliance.

We have emergency procedures in place and emergency response organisations at all our construction sites to both control and limit the impact of incidents and threats.

We periodically audit our operations as well as our business and management practices at all our sites regarding construction health and safety performance and compliance.

Security

Security measures must be integrated into the construction health and safety management system, comply with our construction health and safety policy and meet defined minimal requirements. Construction site security is managed by the local site management and may involve prescreened and prequalified third-party contracted organisations. It is of utmost importance that the logistics security of material supply and storage on site receives the required attention. John Doe Security engages in Customs-Trade Partnership Against Terrorism procedures to comply with site security requirements.

Communication of Risks

We openly communicate and provide information on our construction health and safety performance to all our stakeholders, partners and workers. All communication channels in case of an emergency are defined and functional.

Example 3.2: Another example of a construction H&S policy

CONSTRUCTION HEALTH AND SAFETY POLICY OF ABC CHEMICAL WASTE MANAGEMENT

Introduction

We are a leading waste management organisation in South Africa that adopts better practices for the safe distribution and disposal of chemical waste generated on construction sites and by construction activities.

This document is the construction health and safety policy of the organisation.

Responsibility for Construction Health and Safety

Overall and final responsibility for construction health and safety lies with the board of directors. On a day-to-day basis, responsibility lies with the Health and Safety Manager (HSM), Dr Lubbe, who is duly qualified and responsible for the preparation and review of this policy. The HSM is also responsible for the development and implementation of organisational procedures to ensure safe and healthy working procedures and practices that do not negatively impact on the environment and the clients who we serve as chemical waste managers.

These responsibilities involve:
- Ensuring current health and safety procedures are still effective and relevant to a changing business;
- Transparently passing on information relating to construction health and safety to all our workers; and
- Keeping up to date with legislative and regulatory requirements that relate to construction health and safety.

Construction Health and Safety Policy Statement

We recognise that we have a duty to secure the health, safety and welfare of all our workers by ensuring that the working environment is as safe as is practicable without any threat to their health.

We will fulfil this duty by:
- Ensuring all our workers are competent to do their tasks and are given adequate training, information and supervision; all our vehicles, plant and equipment are properly and regularly maintained; at all times safe handling and use of substances.
- Consultation with workers on matters affecting their health and safety.

This duty extends to those people who visit our premises and to the public in general. These duties and responsibilities extend to all other workers in our plant. It is the duty of every worker to ensure that our safe and healthy working environment can be achieved and maintained by everyone working together, recognising that they are involved in the construction health and safety process and have a duty to secure the well-being of others by observing all construction health and safety regulations and our health and safety policy. To achieve this goal, good communication with respect to construction H&S matters is necessary. Behaviour which puts others at risk will result in appropriate disciplinary action being taken, given our zero tolerance stance on health and safety.

We will make every effort to see that our activities will not harm the environment and will do all in our power to minimise the environmental impact of our operations. To this end we will, whenever possible, attempt to recycle material responsibly, rather than dispose of them. The remaining options, in order of priority, are chemical treatment, with the intention of producing a less hazardous product; high temperature incineration; and landfill. We will reduce the amount of waste that goes to landfill on an ongoing basis and we will comply with all prescribed legislative and regulatory requirements. We recognise that we have a duty of care under the South African environmental legislation for any waste that we consign and we have a procedure for auditing any disposal site or transport contractor that we use for this purpose.

→

Procedure

We have written safe working procedures (SWPs) in place and all our workers receive these procedures in the form of a training pack. Procedures are reviewed and updated biannually or when working conditions change significantly, whichever comes first.

Premises where hazardous substances are stored must comply with the site licence and must:
- Always be kept clean and tidy;
- Contain appropriate and visible warning signs;
- Contain safety and emergency equipment, including personal protective equipment (PPE), spillage kits, fire extinguishers and first aid equipment to comply with construction health and safety regulations; and
- Provide adequate segregation of hazardous materials.

Duties of Workers

It is the duty of every one of our workers while at work to take reasonable care for the health and safety of themselves and of other persons who may be affected by their acts or omissions at work. It is also the duty of every one of our workers to cooperate with any duty or requirement imposed on us by any regulatory organisation. Our workers must not interfere with anything provided to safeguard their health and safety. Any health and safety concerns must be reported to the HSM, Dr Lubbe, or a director of our organisation.

Emergency Procedures and Emergency Equipment

We have emergency procedures in place, which are tested and reviewed periodically. All our workers are made aware of these procedures as part of their training. Any incidents will be recorded and, if necessary, reported to the appropriate authority. We will ensure the adequate provision of emergency equipment and will ensure that equipment will be tested or inspected periodically, if applicable. Fire extinguishers are serviced annually by an independent and qualified third party. Workers must inform the organisation if any equipment is damaged or unusable. The HSM, Dr Lubbe, will make periodic visual checks of safety equipment in the transfer station.

Accidents and First Aid

We have persons qualified in first aid. There are first aid kits in the main office, in the transfer station and in all our vehicles. The transfer station and vehicles also carry eyewash. The transfer station has an emergency shower. The HSM, Dr Lubbe, will make periodic checks of the first aid stock. Any accidents and cases of work-related ill health are recorded in the accident book.

Work Equipment

We will ensure adequate provision of PPE after following the hierarchy of safety management prescribed by the National Institute for Occupational Safety and Health (NIOSH) and will ensure that equipment will be tested or inspected periodically, if applicable. The HSM, Dr Lubbe, will make periodic checks of respirator masks and cartridges in the transfer station. A record is kept of PPE issued to our workers. Workers must inform us when equipment is damaged or unusable. Our directors will be jointly responsible for identifying all plant and equipment needing maintenance. Our directors will be responsible for ensuring effective maintenance procedures and the suitability of new equipment purchased.

Risk Assessments and Method Statements

We realise that we have a responsibility to carry out risk assessments of our activities, including the use of equipment, buildings and hazardous substances. These assessments will be carried out by the HSM, Dr Lubbe, in consultation with workers. The findings of the assessments will be presented to the directors and made known to all affected workers. Our directors will be responsible for ensuring that actions identified in the assessments are implemented. We have written method statements for our activities and these can be provided to any customer who may request them.

$\rightarrow$

Training
The future success of our business is directly dependent on the people who make up the business. No matter what position an individual holds within our organisation, they will have an effect, for good or for bad, on our performance in the marketplace. It is essential that everyone has the skills necessary to carry out their tasks safely and effectively. To facilitate this goal, all our workers will receive job-specific training and a written record of all training is kept, so that it may be referred to in future. It is vital that we ensure that every member of the team is equipped with the right skills and knowledge.

Welfare Facilities
We will ensure that adequate welfare facilities are available at all our sites applicable to the level of staff on the sites.

Contractors and Visitors
All contractors and visitors will be issued with the site rules to make them aware of the emergency procedures and the existence of hazards in the immediate environment on any given day and at any given time.

Review of Construction Health and Safety Policy
This policy, and the organisational procedures to which the policy refers, will be reviewed every two years or when working conditions change, whichever occurs first, and will be distributed to all workers as part of induction and training.

The Directors
21 May 2022

8. Review Questions

- Does the Occupational Health and Safety Act require a formal, written H&S policy?
- What should be included in a construction H&S policy?
- What are the benefits of a construction H&S policy?

9. Review Exercises

- Compile a short construction H&S policy statement for a painting contractor.
- Develop a construction H&S policy for a scaffolding hire and erection company.

Bibliography

Goetsch, DL. 2013. *Construction safety & health.* Upper Saddle River: Pearson Education, Inc.

Griffith, A & Howarth, T. 2014. *Construction health and safety management.* Upper Saddle River: Pearson Education, Inc.

Hinze, J. 2006. *Construction safety.* Gainesville: Jimmie Hinze.

Holt, ASJ. 2005. *Principles of construction safety.* Osney Mead, Oxford: Blackwell Science Ltd.

National Association of Home Builders. 2007. *Home builders' safety program.* Washington: BuilderBooks.

CHAPTER 4

Designing for Construction Health and Safety

John J Smallwood

1. Introduction

Several organisations and many individuals globally have highlighted the role of designers in construction health and safety (H&S), in addition to the role of clients therein.

The International Labour Organization (ILO) states that designers should:

- receive training with respect to H&S
- integrate the H&S of construction workers into the design and planning process
- not include anything in a design that would necessitate the use of dangerous structural or other procedures, or hazardous materials that could be avoided by design modifications or by substitute materials
- consider the H&S of workers during subsequent maintenance.

Within the context of Europe, the Council Directive of 24 June 1992 on the implementation of minimum safety and health requirements at temporary or mobile construction sites highlighted the need for and the role of clients and designers in construction H&S. Each European country then evolved appropriate legislation based on the Council Directive, and the United Kingdom (UK) promulgated the Construction (Design and Management) Regulations in 1994, which entrenched client and designer participation in construction H&S.

The South African Construction Regulations, 2003 followed the direction set by the Construction (Design and Management) Regulations by including clients and designers in the construction H&S process.

Health and safety through design is a fundamental principle of H&S and occupational hygienists invoke the primacy of engineering controls in the hierarchy of controls that are fundamental to the process of hazard reduction. Although architects and engineers regularly address H&S in their designs, they do so with a significant limitation, namely that the concerns apply almost exclusively to the end user of a facility, rather than to the workers who undertake the construction thereof.

Given that legislation was promulgated to improve the energy performance of buildings and structures and promote inclusive design, and reduce their impact on the environment, it makes sense to have done so relative to construction H&S.

Designs develop from Stage 1, project initiation and briefing, courtesy of the client and evolve during Stage 2, concept and feasibility, and further during Stage 3, design development. During each of these stages, a range of designers are involved in the design process and, therefore, have the potential to contribute to mitigating a range of possible design-originated hazards and risks, and introducing construction hazards and risks that will manifest later during the construction process. Furthermore, along with the client brief, architectural designers set the scene in terms of the concept design and design development, which will require other designers to design within constraints. Designs, details, schedules and specifications dictate the materials and methods, and plant and equipment to be used that may entail unnecessary exposure to hazards and risk, which could have been averted through design hazard identification and risk assessment (HIRA). However, designing for construction H&S, prevention through design and design HIRAs will all contribute to mitigating accidents or exposures that result from a combination of circumstances and factors, which in turn, may be traced to several stakeholders. Therefore, they have been included in the Construction Regulations.

2. Legislation

The Occupational Health and Safety Act 85 of 1993 (OHSA) schedules a range of requirements for employers and employees. Therefore, clients and designers must be aware, sensitive and educated relative to construction H&S, particularly given that they visit construction projects and are personally exposed to hazards and risk when doing so.

However, contrary to popular belief, within the context of South Africa, designers have been liable for construction H&S in terms of section 10 of OHSA since 1993, not just since 2003 upon the promulgation of the first version of the Construction Regulations. Section 10, 'General duties of manufacturers and others regarding articles and substances for use at work' of OHSA is extensive in terms of requirements:

- designers, manufacturers, importers, sellers or suppliers are required to ensure, as far as is reasonably practicable, that an article is safe and without risks to health when properly used
- erectors or installers of an article for use at work are required to ensure, as far as is reasonably practicable, that an article is safe or does not create a risk to health when properly used
- manufacturers, importers, sellers or suppliers of any substance for use at work are required to ensure as far as is reasonably practicable, that the substance is

safe and without risks to health when properly used, and provide the necessary information regarding the use of the substance at work, the H&S-related risks when used and the procedures to be followed should an accident involve the substance.

Examples relative to the use phase include cat ladders, lift installations and permanent gondolas for cleaning facades of buildings. These requirements also influence manufacturers in terms of the supply of materials and plant and equipment during the construction stage.

The Construction Regulations of 2014 define a **designer** as:

(a) competent person who:
- prepares a design
- checks and approves a design
- arranges for a person at work under their control to prepare a design, including an employee of that person where they are the employer, or designs temporary work, including its components

(b) an architect or engineer contributing to, or having overall responsibility for a design

(c) a building services engineer designing details for fixed plant

(d) a surveyor specifying articles or drawing up specifications

(e) a contractor carrying out design work as part of a design and build project

(f) an interior designer, shop fitter or landscape architect.

Although the definition is wide ranging and includes several primary project stakeholders, it is largely a copy of the definition included in the UK's Construction (Design and Management) Regulations and should have included construction project managers (CPMs), given that they manage design delivery and the procurement process, and oversee the construction process. The scope of work for CPMs, as set out in the Scope of Work for Categories of Registration for the Profession Regulated by the South African Council for the Architectural Profession, 2019 (Scope of Work) reinforces the latter contention in terms of the interventions and deliverables required of CPMs across the six project stages.

In terms of the Scope of Work, which is regulated by the South African Council for the Architectural Profession (SACAP), it is notable that no H&S services are scheduled for any of the six stages of projects.

The Construction Regulations, 2014 require a range of interventions by clients and designers, in addition to contractors and other stakeholders. Regulation 6, 'Duties of designers', must not be read in isolation as client-related requirements further inform the interventions. Regulation 5(1) requires clients to:

(a) prepare a baseline risk assessment (BRA)

(b) prepare an H&S specification based on the BRA

(c) provide the designers with the H&S specification

(d) ensure that the designer takes the H&S specification into account during design

(e) include the H&S specification in the tender documents.

In terms of Regulation 6, 'Duties of designer', in 6(1) designers are required to:

(a) ensure that the H&S standards incorporated into the regulations are complied with in the design. This is self-explanatory, however, elimination or substitution of materials containing hazardous chemical substances (HCSs) is an example

(b) take the H&S specification into consideration. The H&S specification may require or state that:
- prefabrication should be optimised
- on-site welding is not permitted
- masonry units should not weigh more than x kg

(c) include in a report to the client before tender stage all relevant H&S information about the design that may affect the pricing of the work, the geotechnical-science aspects of the project site and the loading that the completed structure is designed to withstand. Examples of H&S information include the removal of asbestos containing materials, the use of epoxy adhesives, grouts, and resins and interventions related to the temporary support of precast plank/rib and hollow block composite slabs, which ideally would be accompanied by design and construction method statements to ensure integrity of the permanent structure. Geotechnical-science aspects include the findings of trial hole tests and recommended interventions to inform the cost engineer/quantity surveyor in terms of provision for shoring of excavations, if necessary, in the bills of quantity and contaminated land. The loading that structures are designed to withstand is important because of back propping and, as contractors may load recently cast suspended reinforced concrete slabs with materials, plant or equipment, or a combination thereof, this designed loading may be exceeded

(d) inform the client of any known or anticipated dangers or hazards relating to the construction work and make available all relevant information required for the safe execution of the work upon being designed or when the design is changed. The examples included relative to (c) refer

(e) modify the design or make use of substitute materials where the design necessitates the use of dangerous procedures or materials hazardous to H&S. Examples include the substitution of solvent-based paints with water-based paints. However, HIRA is a necessary prerequisite in terms of this requirement. Furthermore, although the raw risk may have been addressed, residual risk may remain. An example includes precast concrete kerbing, which entails the moving, handling and positioning of heavy standard precast concrete kerb sections. The substituting of this form of kerbing with rolled in situ concrete kerbing, which is common in South Africa, introduces ergonomic hazards in the form of bending and twisting, adopting uncomfortable postures, repetitive movements and more, which constitutes the residual risk. An alternative is extruded in situ concrete kerbing, as depicted in Figure 4.1, which, among others, requires bending and a degree of twisting

Figure 4.1: In situ concrete kerb extruding, Las Vegas, USA (Smallwood, August 2015)

(f) consider hazards relating to subsequent maintenance of the structure and make provision in the design for that work to be performed to minimise the risk. An example of this includes a height-compliant reinforced concrete parapet wall to the top of a building with a waterproofed flat roof to falls, constructed to protect maintenance workers when re-coating the waterproofing membrane. A further example includes the incorporation of a permanent gondola, as opposed to the use of a temporary suspended scaffold platform, for the purposes of cleaning the façade and/or fenestration of the structure

(g) when mandated by the client, conduct inspections to ensure conformance of construction to design. If not mandated, then the client's agent is responsible. Examples include the checking and approval of reinforcing steel to suspended reinforced concrete slabs by a structural engineer before casting the concrete

(h) when mandated by the client, stop construction work that is not in accordance with the design's H&S aspects. If not mandated, then the client's agent is responsible. Should the reinforcing steel referred to in (g) not conform to the design, then the related construction activities must cease or not commence until the necessary remedial work has been undertaken, inspected and approved

(i) when mandated by the client, as per (g), during the final inspection of the structure in accordance with the National Building Regulations, 1990, include the H&S aspects of the structure, declare the structure safe for use and issue a completion certificate. This intervention is in the main a closing out of a series of inspections, which is reliant on the conformance of the related work. However, the correct installation of wind bracing relative to prefabricated timber roof trusses is a further example

(j) consider ergonomic design principles to minimise ergonomic-related hazards during all phases of the life cycle of a structure. Examples were presented earlier in (e) and (f).

Regulation 6(2) requires designers of temporary works to ensure that:
(a) they are adequately designed to support all vertical and lateral loads
(b) the designs are done with close reference to the structural design drawings and, if uncertain, consult the contractor
(c) all drawings and calculations pertaining to the design of temporary works are kept at their office and are available for review by inspectors
(d) the loads caused by the temporary works and any imposed loads are clearly indicated in the design.

These requirements are largely self-explanatory, however, it must be noted and digested that temporary works design is a specialised form of design, and that a permanent works designer is not necessarily qualified to undertake temporary works design. The vertical loads are far ranging and do not only include workers, concrete and plant. Furthermore, wind loads, which are lateral loads, can be substantial.

Regulation 6(2)(b) requires consultation of the contractor if uncertainty prevails, which is pertinent because the contractor must undertake the work; however, there may be a need to consult the permanent works designer and reference should have been made thereto.

A notable South African case study is the Bloukrans bridge, which is one of three Garden Route bridges constructed by a consortium of Murray & Roberts and Concor between February 1980 and June 1983. The bridge, which stands at a height of 216 m above the Bloukrans river, has a central span of 272 m and is 451 m in length. The span was constructed using the stayed cantilever method, which involved the erection of a tower on either side of the gorge from which cables temporarily radiated out to support individual sections of the arch until the two halves of the arch were joined in the middle. According to Lambert (2010): 'A tremendous amount of work went into the design of the temporary stage structure which was more complex than the permanent stage'. Two further quotations included courtesy of Steele (1983) reinforce the need for consultation and collaboration between designers and contractors, the need for consideration of the execution of a design, the need for constructability and the implications of the permanent works design for the temporary works design. First: 'notable for the close cooperation and team effort which were achieved by the consultant and contractor, and encouragement given by the client'. Secondly: 'consulting engineers had clearly indicated in their design how the task should be tackled and worked closely with the contractors in converting the drawings they had supplied to reality ...'.

3. Causes of Accidents

The causes of accidents provide insights into the potential contribution of designers to construction H&S in terms of mitigating accidents and exposures. Table 4.1 indicates the causes of accidents relative to clients in the form of contractors insured by The Federated Employers Mutual Assurance Company (RF) (Pty) Ltd (FEM) for 2021, which equates to approximately 50 per cent of the workforce in the South African construction industry. In terms of accidents in general, 'struck by' (32.9 per cent) predominates, followed by 'slip or over-exertion' (13.7 per cent) and 'striking against' (11.1 per cent). 'Struck by' suggests workers being struck by plant or vehicles or falling objects, which, in turn, indicates the potential role of designers. The design and related interventions should, thus, limit the requirement for such plant and vehicles, and the review of H&S plans should interrogate the segregation of plant or vehicles and workers, and the interventions to mitigate falling materials, tools and equipment. 'Slip or overexertion' indicates the need to reduce the physical nature of the construction process, which can be achieved by reducing the need to handle heavy materials and manual handling in general.

In terms of fatalities, 'motor vehicle accidents' (21.1 per cent) predominate, followed by 'inhalation/absorption/ingestion' (19.3 per cent), 'fall on to different levels' (17.5 per cent) and 'struck by' (17.5 per cent). Motor vehicle accidents occur on roads when managers, supervisors and workers are heading to and from construction sites. Although designers cannot intervene directly, they can and should intervene in the H&S specification by highlighting requirements relative to travel and transportation of workers, if they are appointed as the principal agent. 'Inhalation/absorption/ingestion' refers to hazardous chemical substances (HCSs), which constitute potential opportunity for designers to intervene in the form of elimination of the hazardous material or substance with a less hazardous material or substance. 'Fall on to different levels' can be mitigated by designers by reducing the amount of work required at heights. 'Struck by' was addressed relative to accidents in the preceding paragraph.

Table 4.1: Causes of accidents relative to contractors insured by FEM for the year 2021

Cause	Accidents (n)	Accidents (%)	Fatalities (n)	Fatalities (%)	Lost Days (n)	Lost days (%)	Permanent disabilities (n)	Permanent disabilities (%)
Accident type not elsewhere classified	127	1.9	6	10.5	99	0.8	11	1.6
Awaiting information	9	0.1	0	0.0	0	0.0	0	0.0
Caught in, on or between	405	6.1	3	5.3	995	7.8	78	11.0
Contact with electric current	39	0.6	4	7.0	152	1.2	11	1.6
Contact with temperature extremes	110	1.6	0	0.0	460	3.6	40	5.7
Fall on to different levels	594	8.9	10	17.5	2652	20.8	131	18.5
Fall on to same level	324	4.9	0	0.0	250	2.0	42	5.9
Inhalation/absorption/ingestion	657	9.8	11	19.3	328	2.6	5	0.7
Motor vehicle accident	531	8.0	12	21.1	1399	11.0	56	7.9
Slip or overexertion	918	13.7	0	0.0	1369	10.7	50	7.1
Striking against	743	11.1	0	0.0	970	7.6	64	9.1
Struck by	2 199	32.9	10	17.5	4032	31.6	216	30.6
Unclassified – insufficient data	21	0.3	1	1.8	47	0.4	3	0.4
Total for 2021	6 677	100.0	57	100.0	12753	100.0	707	100.0

4. Designing for Construction H&S and the Impact thereon

There is no more important stage in the construction process than that of the design stage because during this stage conceptual ideas are converted into constructable realities. Designing for H&S is one of the designing for constructability principles. A variety of considerations need to be balanced simultaneously that include, among others, designing for H&S. Designing for H&S is an integral part of the wider design process, which, therefore, needs to be included in design planning because doing so will result in healthier and safer construction and maintenance of structures and facilities.

In terms of the impact of design on construction H&S, 450 reports of construction workers' deaths and disabling injuries in the United States of America (USA) were analysed to determine whether addressing H&S in the project designs could have prevented the incidents. The findings of this research determined that in 151 cases (33.6 per cent) the hazard that contributed to the incident could have been eliminated or reduced if design for H&S measures had been implemented.

5. Occasions Designers should Address or Consider H&S

There are several opportunities and occasions during the construction process for designers to consider H&S.

5.1 Client meetings

Client meetings commence with the client briefing and designers may, depending on their form of appointment, lead the process in their capacity as the principal agent, or they may be part of the briefing process. Clients should be informed or reminded of their responsibilities in terms of the OHSA, and the Construction Regulations, 2014. Clients may have H&S requirements, objectives and goals in terms of undertaking projects, which should be ascertained. Clients should be informed of the H&S implications of decisions relating to the height of a building, type of structural frame, general design, materials, procurement system, procurement process, type of contract and project duration.

5.2 Deliberating project duration

The design, procurement and construction processes should be appropriate in terms of their duration. The duration of the construction process should be compatible with the scope, value, volume, number of floors, total area (m^2), range of activities and the sequencing of activities. Furthermore, the construction process may, for example, require the employment of labour from the surrounding communities, in which case, training of this local labour force in general, but also H&S training, will be required.

The project duration should be addressed and confirmed before commencing the concept design. It should be discussed in detail during the client briefing. The design team should be assured that they can complete what is required in terms of the design process during Stage 2, concept design, and Stage 3, design development, and that construction can be completed in terms of the design within the project duration. Louis Sullivan, the legendary Chicago-based architect, stated that 'form follows function'. Therefore, the following should apply relative to project duration: 'project duration follows design'; 'design follows construction'; and 'project duration follows construction'. The traditional approach of compressing the design and construction processes to suit the project duration is counterproductive and has been red flagged by several H&S-related studies.

Recent research conducted in Australia determined the following relative to the effects of a five-day working week on construction workers and their families:

- the majority of workers (75.4 per cent) preferred a five-day work week as opposed to either a six- or seven-day work week
- workers reported an improvement in their work-life balance, 50 per cent cited 'a great difference' and 28 per cent 'some difference'
- next of kin noticed improvements in their partner's mood and well-being during the research project, citing less fatigued, more relaxed and more available to enjoy their social and family life
- next of kin were better able to pursue employment, and enjoy some respite from parenting and domestic responsibilities.

5.3 Three stages of design

Generally, the design process is divided into three stages, namely concept, scheme and detailed. Each design stage addresses different issues, however, the issues evolve from the general, during concept design, to the particular, during detailed design. Concept design addresses project-wide issues and includes major location and operation issues. Scheme design addresses system issues and addresses the major disciplines and interface issues. Detailed design addresses elemental issues such as component details and assembly issues.

5.3.1 Concept design

The concept design constitutes a precedent because the initial concept of the building or structure invariably has H&S implications and provides information to the project client and/or team with respect to the scheme and detailed design. The height and shape of the building or structure on plan, type of structural frame and elevations have a range of H&S implications in terms of, among others, plant and equipment and temporary works required during construction.

In the case of a warehouse, the concept design involves a portal frame structure consisting of stanchions and prefabricated girder roof trusses to limit the amount of work at elevated heights.

5.3.2 Scheme design

The scheme design develops and links the concept and detailed design and, in the case of the warehouse example, the size of the prefabricated girder roof trusses is designed such that they can be hoisted into position by a mobile crane.

5.3.3 Detailed design

The detailed design consolidates and expands the concept design and scheme design during Stage 3, design delivery. In the case of the warehouse example, the detailed design would focus on the girder/bracing and girder/stanchion connections in terms of bolted connections and appropriate tolerances to facilitate assembly while erectors work from elevated working platforms (EWPs).

5.4 Design coordination meetings

Design coordination meetings are critical in terms of constructability, the avoidance of clashes between services and the finished product. However, they are necessary to remind the design team of the impact of its design on H&S and the interventions that can enhance construction H&S and H&S during the maintenance of completed buildings and structures.

Examples of H&S items that could be discussed and resolved during these meetings include the depth of suspended ceiling spaces to accommodate the range of services required and access during installation and maintenance, which constitutes an ergonomics issue. Figure 4.2 illustrates the realities experienced during the installation of services in ceiling spaces.

Figure 4.2: Congested ceiling space, Cape Town (Smallwood, 2004)

5.5 Constructability reviews

There are 12 principles of constructability, several of which have H&S implications. The construction programme should be realistic, construction sensitive and have project team commitment. The resultant construction methodology should be considered during design. The technology of the design should be matched to the skills and resources that are likely to be available during construction. Accessibility can be enhanced if considered during the design and construction stages, for example ducts for services such as water supply pipes and soil pipes. The design should enable the use of innovative techniques during construction such as, for example, flat slabs without beams enable the use of table forms, which mitigate major ergonomics challenges during the erection and striking of support work and formwork.

Constructability reviews should be conducted from Stage 1, project initiation and briefing, and throughout Stage 2, concept and feasibility, and Stage 3, design development. Post-design, during Stage 4, construction documentation and management, and Stage 5, project closeout, including assessment thereof in the project closeout report.

5.6 Preparing project documentation

All project documentation should refer to construction H&S:

- site plans
- architectural floor plans, elevations, sections and details
- structural plans, reinforcement layouts and bending schedules
- civils works
- mechanical installation layouts
- electrical installation layouts
- schedules such as floor coverings, colour and glazing
- landscaping layouts and schedules.

Site plans should indicate:

- contours
- water table
- position of existing and future services
- position of trial holes
- collapsible soils
- contaminated ground
- height of building or structure
- level of basements, if any
- heights of buildings or structures adjoining the site
- direction of traffic to roads adjoining the site
- restrictions in terms of entrances from and exits to such roads.

Architectural floor plans, elevations, sections and details should indicate the position of activities that entail raw risks that cannot be mitigated, and residual risks.

Structural plans should indicate design loadings and special interventions, including reference to design and construction method statements, in the case of cantilever slabs, and suspended composite slabs consisting of precast planks/ribs, hollow blocks and in situ concrete beams and overlay. Reinforcement layouts and bending schedules should indicate raw risks that cannot be mitigated and the residual risks. Examples of the former include long lengths of large diameter reinforcing steel bars.

Civils works plans and drawings should indicate:

- contours
- water table
- position of existing and future services
- position of trial holes
- collapsible soils
- contaminated ground
- raw risks such as:
  - deep excavations
  - heavy pipes in terms of mass per lineal meter
  - kerbing consisting of heavy precast concrete kerb sections
  - hot mix asphalt road surfacing.

Residual risks such as in situ rolled kerbing or extruded in situ kerbing, should also be annotated on such plans and drawings.

Landscaping layouts and schedules should indicate the raw risks that cannot be mitigated and the residual risks. Examples of raw risk include:

- collapsible soils
- high water table
- the approximate height and diameter of the trunk and root system, and mass of large trees such as palms, including any thorns
- hardscaping in the form of heavy timber railway sleepers
- the application of poisonous sprays.

5.7 Pre-tender site meeting

H&S should be an agenda item and discussion should be linked to the contractor H&S specification, which should include both the raw and residual risks, as highlighted in the designer report, and should highlight key issues relative to the future H&S plan. The pre-tender site visit constitutes an ideal opportunity for prospective tenderers to experience the project environment in the form of:

- public thoroughfares
- possible restrictions applicable to say site entrances

- hazards, such as overhead powerlines
- adjacent buildings, which may restrict the use of tower and mobile cranes and other plant.

5.8 Pre-qualification of principal contractors in terms of H&S

Designers may or may not fulfil the function of principal agent. However, regardless of their role in terms of project management, they should be involved in any pre-qualification review process to assist with assessment of the potential PCs approach to managing H&S.

5.9 Evaluating tenders

Regardless of their role in terms of construction project management, designers should be involved in the evaluation of tenders for several reasons, not just for H&S. The assessing of the adequacy of the PC's financial provision for H&S is of particular importance in terms of the addressing of raw and residual risks, which is a function of the design process. This highlights the importance of optimum BRAs, designer H&S specifications, designer reports and contractor H&S specifications.

5.10 Site handover

All designers should be present and contribute, relative to their discipline, in terms of the related H&S issues. Examples include:
- access to the site from public thoroughfares, including restrictions
- constraints in the case of brownfield projects
- health risks arising from the client's activities such as sewage treatment works and nuclear power plants
- services-related issues, including the position of electricity, fire main, gas, internet, sewage and water supply, in the case of brownfield sites, or connection points for such services, in the case of greenfield sites
- an overview of existing structures to be demolished or altered.

5.11 Working drawings

Working drawings, details and schedules issued for construction should be annotated such that they indicate or reference raw risks that could not be mitigated, residual risks and design and construction method statements.

5.12 Site inspections

Designers undertake site inspections to ensure, among others, that work conforms to design. However, during site inspections, designers should also review the extent to which design and construction method statements have been complied with; the impact of their design on construction H&S; and identify cases where they could have reduced the raw risk or for that matter, the residual risk and, more importantly, design originated risks that were not identified.

5.13 Project progress meetings

Designers attend project progress meetings, which discuss project wide issues that include H&S. Issues such as conformance of work to design, design and construction method statements, design originated hazards, concrete cube and/or core test results, tension test results relative to post-tensioning cables, the implications of variation orders and additional work on H&S should be discussed.

5.14 Site meetings

Ad hoc site meetings may be undertaken by designers to discuss issues, challenges or details. H&S should always be borne in mind and considered or referred to. Such meetings also constitute a further opportunity for designers to review the impact of their design on construction H&S, identify cases where they could have reduced the raw risk or, for that matter, the residual risk and, more importantly, design originated risks that were not identified.

6. Aspects of the Project Design that should be Considered in Terms of H&S

The aspects of the project design that should be considered in terms of H&S further amplify the need for designing for construction H&S and design HIRA. Given the advent of digitalisation in the form of, among others, building information modelling, the consideration of construction H&S during the design process, the identification of hazards and the assessment of the related risk has been facilitated and simplified.

6.1 Site location

Although clients decide, or influence the decision, on the location of a project, designers may recommend a location. The location of a project has implications in terms of:

- the weather in the form of wind, precipitation, air temperature and lightning
- socioeconomic issues, such as community involvement in projects, crime and vandalism

- activities undertaken on adjoining properties
- the potential to contract malaria and encounter wild animals, such as crocodiles and snakes.

6.2 Features and general design

The features and general design of a building or structure, which are, to a large extent, dependent upon the client's requirements, influence the site location. However, the site location, in turn, influences the features and general design. Furthermore, the site location is influenced by the required geotechnical requirements relative to the features and general design of a building or structure.

However, the general design of a building or structure has a profound impact on H&S as a result of factors other than the site location. The Bahá'í Temple in New Delhi, India, depicted in Figure 4.3, is in the shape of a lotus flower and the 27 concrete and marble petals that constitute the structure are divided into three different types, namely nine entrance leaves, which mark the doorways; nine outer leaves, which form the roof space; and nine inner leaves, which enclose the central hall in an interior dome. The construction thereof had major implications in terms of, among others, the support work, formwork and access platforms.

Figure 4.3: Bahá'í Temple, New Delhi, India (Smallwood, 2005)

6.3 Surveys/archival reviews

A range of surveys need to be conducted during Stage 2, concept and feasibility, and possibly during Stage 3, design development. The range is influenced by whether the project is a brownfield or greenfield project. Brownfield projects are challenging because they constitute existing facilities. In theory, an informative H&S file should assist with respect to a brownfield project constructed after the promulgation of the Construction Regulations in 2003.

Geotechnical surveys are critical in terms of the structural design, including the founding of the building or structure, in addition to informing the project team with respect to provisions and interventions to ensure excavation H&S. Contaminated land surveys are necessary because of environmental legislation and regulations, but also because of the H&S hazards and risks posed during earthworks and augured piling, if land is identified as contaminated.

The location of existing services, private and public, is necessary for the permanent works design, in addition to designing for construction H&S.

In the case of brownfield projects involving alterations and/or additions, the integrity of the structure needs to be determined, which should include interrogation of plans, reinforcing layouts and schedules, as archived. It may be necessary to conduct, among others, core tests to verify concrete strength.

Further surveys should address hazardous materials in the form of asbestos and lead; hazardous areas such as confined spaces; and hazardous areas and locations such as fragile roof lights.

Site access and other potential restrictions should be identified from a temporary works perspective, as site entrances should not compromise traffic safety and public H&S.

Local environmental conditions and adjacent land uses should be identified.

In the case of occupied sites, such as manufacturing or process plants, schools and shopping malls, a range of issues need to be considered and details secured. These issues include:

- access control
- permit to work systems
- prohibited activities
- maintaining the integrity of the fire protection system
- non-compromising of emergency procedures.

Fire protection is of particular importance because the related installation must not be compromised during design and/or construction.

6.4 Site coverage

The site coverage or footprint of the building or structure has implications for the construction process and H&S in terms of the site layout during construction, the availability of space for the site establishment, storage and laydown areas, temporary works and the positioning of plant and equipment.

6.5 Plan layout

Regular plan layouts of multistorey buildings are preferable, as irregular or intricate layouts involving, say, indentations or setbacks, create challenges during the construction of the structure, including the erection of scaffolding. Openings in floors should be consistent in terms of position, that is, verticality, which will assist in terms of opening protection.

6.6 Type of structural frame

Within the South African context, there are four types of structural frame:
1. loadbearing masonry
2. timber frame
3. structural steel
4. reinforced concrete.

Each type involves challenges that include H&S and ergonomics.
- **Loadbearing masonry** involves extensive manual handling, bending, twisting and repetitive movements, and requires substantial on-site storage areas. Furthermore, cement mortar is required for bonding the masonry units and cement plaster for rendering, which adds to the on-site storage challenges and exposes workers to cementitious materials, which include chromates courtesy of the cement.
- **Timber frame** construction, in turn, has several advantages in the form of handling light weight prefabricated frames, which entail less onsite work, promote mechanisation and require reduced onsite storage, because they can be directly positioned on the substructure upon offloading. Timber frame construction also involves less manual handling. Furthermore, the attaching of the external fabric and internal skin, in the form of weather boarding or similar, and gypsum sheeting, respectively, is less physically demanding than rendering, such as cement plaster, in the case of masonry construction. Additional advantages include schedule or time benefits, less wastage and reduced impact on the environment. Consequently, in general, they result in less exposure to ergonomic hazards than other types of structural frames.
- **Structural steel** frame structures involve less on-site work, which often features pre-assembly, which results in less manual handling. However, they often involve

very challenging work at elevated heights, work which is ideally undertaken using elevated working platforms where possible. Furthermore, the floor structures tend to follow the frame and, therefore, there is a greater need for fall prevention and protection, which can be facilitated through designing for construction H&S interventions.

- **Reinforced concrete** frame structures involve support work and formwork, steel reinforcing or post-tensioning and in situ concrete, all of which entails major plant and equipment and logistics management. Furthermore, reinforced concrete frame structures result in physically demanding and challenging work at elevated heights. Ergonomic hazards include extensive manual handling, bending and twisting, use of body force, reaching away from the body and repetitive movements.

6.7 Elevations

The nature of elevations in terms of their design, details and the related enclosing fabric, including method of attachment, has a range of H&S implications in terms of access, materials and methods, plant and equipment, in the form of temporary works required, and exposure to hazards and risk, including working at elevated heights.

Figure 4.4 depicts the installation of fenestration to a façade of a building, indicating the implications of design for construction and subsequent maintenance.

Figure 4.4: Externally installed fenestration, Dublin, Ireland (Smallwood, 2004)

6.8 Specifications

Specifications, in general, prescribe construction materials and methods. However, specifications may have negative implications for workers; for example in the case of materials that contain HCSs, such as flooring and epoxy adhesives, grouts and resins. Specifications of herringbone pattern concrete block paving that is bordered with kerbing, edging, channels or a header course edge will require a major amount of cutting with a brick cutting machine or, worse, an angle grinder, which will result in exposure to hand-arm vibration, noise and dust. Basketweave or a parquet pattern would avert such cutting.

6.9 Components and prefabrication

Pre-assembled components, such as pre-glazed aluminium or PVC window and door units, mitigate duplicate manual handling, mitigate fragmentation of production and reduce the number of activities on site. Off-site prefabrication, in the form of pre-assembled structural steel elements and prefabricated timber roof trusses, mitigates manual handling and limits the amount of exposure to working at heights.

6.10 Position of components and fittings

The position of components has implications for H&S during construction and the use phase of the building or structure. An example includes the positioning of light fittings above indoor swimming pools and in difficult to access places during the use phase, as in the case of the porte cochère depicted in Figure 4.5.

Figure 4.5: Position of light fittings beneath porte cochère, Kruger Mpumalanga International Airport, Mbombela, Mpumalanga (Smallwood, 2004)

6.11 Finishes

Finishes are pervasive in building construction and are required to floors, walls, ceilings and other surface areas, such as columns. Finishes should not involve excessive adopting of awkward postures, kneeling, manual handling, reaching overhead, reaching away from the body and repetitive movements. Ceilings, for example, should be self-finished and should not involve further rendering, which also has a positive impact in terms of mitigating maintenance requiring overhead work.

6.12 Surface area of materials

The surface area of materials, such as glass panels for shopfronts, drywall panels and ceiling panels forming part of components or elements, should be considered because of the manual handling-related implications and the loads that need to be supported during the installation and fixing processes.

6.13 Mass and density of materials

The mass and density of materials both have implications in terms of transporting, manual handling, cranage, other forms of hoisting and temporary works, such as scaffolding and support work. Consequently, designers should be sensitive to the mass and density of materials they specify to be used in the construction phase of their designs. The mass of unit materials such as masonry units can be readily determined; however, in the case of the density of materials such as natural stone used for cladding, designers must be conscious and mindful of the resultant mass of panels detailed and specified as a result of their surface area and thickness.

6.14 Content of materials

HCSs are a major issue in construction and, ideally, should be eliminated or substituted in terms of the hierarchy of controls, as opposed to relying on engineering controls, safe working procedures (SWPs) and the use of personal protective equipment (PPE). HCSs enter the body by ingestion (swallowing), inhalation and through the skin. Airborne contaminants exist in the form of dusts, fumes, vapours, mists and gases. Dusts, in turn, are created by processes such as sawing, grinding, sanding, abrasive blasting or when granular materials such, as sand, cement and stone, are mixed, blended or transferred. Inhalation is the most common hazard and small particles can reach deep into the lungs, resulting in pneumoconiosis, asbestosis or lung cancer, depending on the material. Solvents can be inhaled, resulting in, among others, respiratory irritation and damage to the central nervous system as a result of long-term exposure. Welding causes the release of a variety of complex fumes that can cause metal fume fever. Solvents can be absorbed through the skin causing dermatitis, while concrete causes an irritant and allergic contact dermatitis because of its abrasive and alkaline properties. The addition of chromates to cement increases the risk, while slaked lime causes cement burns. Bitumen and related products cause dermatitis and acne type skin diseases.

6.15 Edge of materials

The edges of certain materials may be sharp and, thus, have the potential to cause cuts to the human body, especially to hands and digits during manual handling and installation. Furthermore, the cutting of certain materials may expose workers such hazards.

6.16 Texture of materials

The texture of certain materials may be rough and, thus, have the potential to abrade, say, hands during manual handling and installation. Although the wearing of gloves can afford protection in such cases, PPE is at the end of the hierarchy of controls and, therefore, constitutes a measure of last resort, in that such materials should be eliminated or substituted.

6.17 Details

When detailing, designers should consider the impact on constructability, ergonomics and H&S in terms of, among others, lifting, pushing, pulling, reaching, hand-arm vibration and repetitive movements. Further considerations include accessibility during construction and the use phase of the building or structure.

6.18 Method of fixing

Activities such as the installation of services entail a range of implications and constitute a challenge in terms of, among others, the method of fixing. Furthermore, the type of structural frame to which building services are attached further amplifies the challenge. For example, in the case of suspending air conditioning ducting or cable trays from a composite slab relative to a structural steel framed structure, the suspension hangars can be positioned through the corrugated permanent steel decking before casting the in situ concrete overlay. This mitigates having to drill into the soffit of a slab, as in the case of a reinforced concrete structure, to insert anchors or bushes to receive threaded suspension rods.

6.19 Schedules

Designers compile a range of schedules during the design and construction processes, and several H&S issues related thereto have been addressed in terms of, among others, specifications, content of materials and mass and density of materials. However, in some cases, the compiling of schedules should alert designers to the construction realties of their design such as, for example, the kg mass per metre of, say, a Y32 or T40 reinforcing steel bar. Furthermore, the lengths of reinforcing steel bars should be considered relative to the reinforced concrete element and the fixing thereof while, for example, sliding a reinforced concrete lift shaft core or silo because the working platform may constitute a constraint in terms of accommodating the length of the bars.

7. Review Questions

- Was the inclusion of designer responsibilities in the Construction Regulations justified? A philosophical and pragmatic response is required.
- To what extent does the H&S inspectorate of the DEL interrogate designer compliance with the designer-related requirements of the Construction Regulations? A pragmatic response is required.
- What aspects of designer contributions to construction H&S can be improved?

8. Review Exercises

- Compile a designing for construction H&S checklist for all design disciplines for reference to by a CPM or principal agent.
- Compile a 750-word article titled *The holistic role of designers in construction H&S*.

Bibliography

Behm, M. 2006. *An analysis of construction accidents from a design perspective.* Silver Spring: The Center to Protect Workers' Rights.

Commission of the European Communities. 1993. *Four guides for the 'temporary or mobile construction sites' directive.* Brussels: Office for Official Publications of the European Communities.

Construction Industry Development Board (cidb). 2009. *Construction health & safety in South Africa: Status & recommendations.* Pretoria: cidb.

Hecker, SF, Gambatese, JA & Weinstein, M. 2006. Designing for construction safety in the US: Progress, needs, and future directions. In: RN Pikaar, EAP Koningsveld & PJM Settels (eds) *Proceedings of the 16th Triennial Congress of the International Ergonomics Association*, Maastricht, Netherlands (10–14 July 2006).

International Labour Organization (ILO). 1992. *Safety and health in construction*. Geneva: ILO.

Lambert, L. 2010. Our legacy. *Robust*, September: 13.

Steele, D. 1983. Bloukrans bridge. *Concrete Beton*, 30(06): 10–11.

The Australian Human Rights Institute. 2021. *Project 5: A weekend for every worker*. Sydney: The Australian Human Rights Institute, University of New South Wales.

The Federated Employers Mutual Assurance Company (RF) (Pty) Ltd (FEM). 2022. *Statistics by injuries*. Johannesburg: FEM, https://roe.fem.co.za/Stats#/Accident-Stats.

Thorpe, B. 2006. *Health and safety in construction design*. Aldershot: Gower Publishing Limited.

CHAPTER 5

Baseline Risk Assessment

John J Smallwood

1. Introduction

In terms of Construction Regulation, 2014 Regulation 5(1)(a), clients are required to prepare a baseline risk assessment (BRA) for an intended construction work project. The Constructions Regulations do not define a BRA and neither do they provide information about the focus thereof, recommended sections, contents or stage of the project when it should be conducted.

BRAs are one of three types of risk assessments and are intended to determine the wide range of risks associated with a project and the status thereof, resulting in a risk profile for a project. BRAs are not issue-based or continuous risk assessments. However, they are overarching in nature and include the:

- risks the client faces in terms of the project concept and during the project phases
- risks, including health and safety (H&S) risks, that the client and designers face in and from the project design
- residual risks the client and designers have not resolved that influence construction H&S and user phase H&S.

2. Rationale and Intention

In terms of Construction Regulation 5(1)(b), clients are required to prepare an H&S specification based on the BRA, which constitutes the legal rationale; however, the non-legal rationale for undertaking a BRA is wide ranging. The construction industry is client-driven, which was the rationale for the inclusion of client responsibilities in the 2003 version of the Construction Regulations, which were expanded in the 2014 version.

Most clients evolve the concept and have a vision of the proposed project, be it a brownfield or a greenfield project. Furthermore, they may have an understanding and an appreciation of the project requirements and, more importantly, the implications thereof. Consider the refurbishment of a shopping centre. The client should be aware of the plethora of aspects and issues relative thereto, among others:

- number of tenants
- square meterage of shops and trafficable areas
- low, shoulder and peak shopping seasons
- lowest, mean and highest number of shoppers per day
- position of services and emergency procedures.

Figure 5.1 shows some of the interventions in the form of overhead and related protection during the Greenacres Shopping Centre redevelopment project, Gqeberha, to enable safe shopping during the project.

Figure 5.1: Overhead and related protection during the Greenacres Shopping Centre redevelopment project, Gqeberha (Smallwood, 2016)

Furthermore, there is the issue of the business case for the project, including the feasibility of executing the plan. A further aspect to be considered is the shareholders, in the case of a public or private company, or other business form, which, along with the financier(s), should be assured that possible risks have been identified, quantified, assessed and responded to.

Historically, South Africa has experienced a series of shopping centre collapses. On 20 December 2001, the Kolonnade Shopping Centre's concrete roof collapsed onto the ice skating rink, injuring 21 people. This collapse highlighted the notion that the outcome of collapses or failures is fortuitous. Although 21 injuries are 21 too many, miraculously there were no fatalities. The negative publicity in the media from such collapses, regardless of the allocation of responsibility, is not in a client's, construction H&S agents' (CHSA), construction project manager's (CPM), designer's or contractor's best interests. Therefore, when undertaking brownfield projects and, to a lesser extent, greenfield projects, clients are at substantial risk and it is in their commercial interest to be circumspect and to adopt a scientific approach when deliberating the undertaking of such projects.

The Table Mountain Aerial Cableway upgrade project, undertaken in 1997, constituted a challenge because Table Mountain, at a height of 1 086 m, is home to 1 460 different species of fynbos, and forms part of the Cape Floral Kingdom. In 2000 the mountain was declared a United Nations Educational, Scientific and Cultural Organization (UNESCO) World Heritage site. The flora was merely one challenge; the other challenges being the fauna, the weather and the location. Although the project was completed prior to the implementation of the Construction Regulations in 2003 and the amended Construction Regulations in 2014, the latter requiring a BRA, it constitutes an ideal example of why BRAs are required. Figure 5.2 illustrates the nature of and challenges posed by the project.

Figure 5.2: Upper Aerial Cableway Station, Table Mountain (Table Mountain Aerial Cableway, 2014)

The Strand Street helicopter crash on to the top of a building in Cape Town in the Western Cape on 10 February 1999 resulted in the death of the four occupants of the helicopter. The helicopter was conveying an underslung load in the form of an air conditioning unit, to the roof of a building when its tail rotor hit a roof-mounted billboard and the tail boom sheared off. The pilot lost control and the helicopter fell onto the roof of an office block, exploding on impact. Although the South African Civil Aviation Authority examined the operator's pre-mission planning and reconnaissance, the marshalling arrangements and the weather, this accident constitutes a further example of why BRAs are imperative in all projects. Figure 5.3 shows the crashed helicopter in flames on the rooftop of the building.

Figure 5.3: Crashed helicopter in flames on rooftop of building, Strand Street, Cape Town (FlightGlobal, 1999)

The importance and relevance of nature and the weather to projects is underscored by the death of six construction workers when the structure they were sheltering under was struck by lightning on 18 March 2015, while a further five workers were injured in the incident. The workers had been working along the N8 highway outside Botshabelo near Bloemfontein in the Free State, when it started raining and the men took shelter under an unidentified structure when the lightning struck.

The section 32 inquiry into the Injaka bridge collapse in Bushbuckridge, Mpumalanga on 6 July 1998 resulted in a comprehensive report that, among others, highlighted the importance of assuring the appointment of competent designers. In terms of section 4, 'Cause of the Incident', '4.1. It was established from the evidence gathered that', among others, '4.1.8 The use of design and construction personnel, at decision-making level, without appropriate qualification and experience in incremental launched bridges, especially ...' was one of several major direct causes of the accident. This indicates a further need for BRAs to be project wide, and not issue-based or continuous risk assessments, and to address issues that impact on H&S indirectly.

3. Who should Undertake the BRA?

Clients may appoint a CHSA to fulfil their responsibilities in terms of the Construction Regulations, a CPM or principal agent to lead the project through the six stages, the designers, a quantity surveyor (QS) or cost engineer, the principal contractor (PC) and the nominated subcontractors. Therefore, the client should understand and appreciate the nature, scope and complexity of the proposed project to ensure informed appointments of the aforementioned. This, in turn, requires that those appointed have the requisite competencies to deal with the challenges. Surface competencies in the form of knowledge and skills are one group of competencies, which, in terms of conducting BRAs, include a range of knowledge areas and skills. Knowledge areas include, among others:

- mind mapping
- system thinking
- scenario planning
- construction project management
- design
- construction management
- construction materials and methods
- plant and equipment
- temporary works
- risk management
- H&S
- risk assessment.

Skills include, among others:
- mind mapping
- system thinking
- scenario planning
- conceptualising
- planning
- forecasting

- programming
- scheduling
- risk assessment.

However, a second group of competencies, namely core competencies, is critical because these competencies contribute to differentiating between above average and average performance, the latter resulting from merely possessing the surface competencies. The 10 core competencies in three categories relative to their respective three categories are:

1. **self-concept**: values, aptitude, attitude and self-image
2. **traits**: self-confidence, team player and able to handle ambiguity
3. **motives**: focus on client success and preserve organisation/personal integrity.

Although all 10 core competencies are relevant to H&S, and to project performance in general, it is imperative that:

- H&S be included as a project value not as a priority
- clients and stakeholders have the right attitude and the aptitude for challenging projects, are team players and can handle or deal with ambiguity
- stakeholders focus on client success and preserving the client's integrity in addition to their organisation's integrity.

4. When should the Baseline Risk Assessment be Conducted?

As stated, the Construction Regulations are silent with respect to when the BRA should be conducted. However, the scope of services for CHSAs, as set out in The Scope of Work for Categories of Registration of the Project and Construction Management Professions, 2019 records: '2.6 Prepare draft construction project baseline risk assessment' relative to Stage 2, concept and feasibility. The question arises: What if the BRA indicates several to many extreme risks at Stage 2? This question is reinforced by the following included relative to Stage 1, project initiation and briefing, in the scope of services for CHSAs as they require outputs from the BRA:

1.2   Assist in developing a clear construction project health and safety brief.

1.5   Advise on the necessary surveys, analyses, tests and site or other investigations where such information will be required for the next stage of the project.

1.7   Identify construction project health and safety risk profile.

1.8   Provide necessary information within the agreed scope of the construction project to the other consultants.

Clearly, the BRA needs to be conducted during the infancy of Stage 1 to provide the outputs required for the other services rendered by CHSAs to the client, CPM and designers.

Although it can be argued that hazards and the related risks should be progressively addressed, prioritisation thereof through quantification in the form of a risk rating can provide an agenda for addressing the major hazards and risks of a project.

5. Focus of the BRA

A study reviewed 100 construction accidents in the United Kingdom (UK). The first category includes 'immediate accident circumstances and shaping factors', namely:

- worker and work team factors
- workplace factors
- materials and equipment.

The second category includes 'originating influences', namely:

- economic influences
- client requirements
- H&S education and training
- construction design and processes
- management of the project site
- construction processes
- H&S culture
- risk management.

In terms of BRAs specifically, the following categories of risks relative to the acronym PEPMELF: people; equipment; process/procedure; materials; environment; legal, including liability; and financial.

In terms of risk management relative to organisations, HM Treasury in the UK refers to six external categories of risk, which, although they arise from the external environment and, therefore, are not wholly within an organisation's control, exposure to the risk(s) can be mitigated. The six categories that constitute the acronym PESTLE are political, economic, sociocultural, technological, legal/regulatory and environmental.

The next risk management process referred to by HM Treasury is relative to operational risks, which are those risks relating to existing operations, which in turn, include those risks relating to current delivery and building and maintaining capacity and capability:

- **delivery**: service/product failure and project delivery
- **capacity and capability**: resources, relationships, operation and reputation

- **risk management performance and capability**: governance, scanning, resilience and security.

The next process is relative to change, which includes risks created by decisions to pursue new endeavours beyond the organisation's current capability and includes the categories:

- probabilistic safety assessment targets
- change programmes
- new projects
- new policies.

The six PESTLE external categories of risk, operational risks and change-related risks, provide insights into to what BRAs should address, especially in light of the fact that construction H&S is a multi-stakeholder issue and that H&S performance is affected by design, procurement and construction. Furthermore, the relationship between H&S and the other project parameters, such as cost, environment, productivity, quality, productivity and time, further support the need for a risk assessment.

The BRA should include anything that influences and impacts on construction H&S indirectly and directly. The indirect category will generally include issues at national, industry, organisation and project level, while the direct category will generally be project-wide (level) hazards per se.

An indirect category example would include a national and a construction economy experiencing a boom period, which will impact on the availability of, among others, skilled managers, supervisors and production workers; plant and equipment; and materials. The 'construction mafia' and small, medium and micro enterprises (SMMEs) constitute issues at industry and project level. Further project-wide issues include:

- project duration
- client briefing
- procurement system
- contract documentation
- appointment of the CPM, CHSA, designers, a cost consultant and contractors
- management of information.

The direct category will generally include project-wide hazards and threats, among others:

- socioeconomic
- local communities
- heritage issues
- access to the proposed site
- topography of proposed site

- ground conditions
- existing services
- availability of services
- weather
- fauna and flora
- viruses and pandemics
- position of adjoining buildings and structures
- activities adjacent to or beneath the site.

The indirect and direct categories are further elaborated upon.

The following are briefly discussed relative to the indirect category:

- client briefing such as, for example, the implications and/or the impact of an inadequate client briefing
- project duration such as, for example, the implications and/or the impact of a compressed project schedule in terms of working hours per day and working days per week, and the impact thereof on management, supervision and workers' H&S, as a result of fatigue, and their well-being because of a lack of time to rest and recuperate, and attend to family-related responsibilities
- procurement system such as, for example, the implications and/or the impact of an inappropriate procurement system on H&S
- conditions of contract such as, for example, the implications and/or the impact of inappropriate conditions of contract that do not make adequate reference to H&S
- digitalisation such as, for example, the implications and/or the impact on H&S of not digitalising by adopting, say, building information modelling to facilitate designing for H&S and design HIRA
- access to the project location such as, for example, infrastructure such as airports, harbours, railway routes above and below ground, rivers and roads, including conditions, load restrictions and restrictions with respect to transportation of specific goods
- 'construction mafia' such as, for example, extortion and physical violence
- SMMEs such as, for example, extortion, forced project participation, competencies and resources
- appointment of project stakeholders. There are a range of aspects/issues (hazards), implications (consequences/severity) and probability relative to each of:
  - CPM or principal agent, CHSA and designers such as, for example, fees, competencies and resources
  - PC such as, for example, tender price, competencies, resources and financial provision for construction H&S
  - nominated and PC appointment of subcontractors such as, for example, tender price, competencies, resources and financial provision for construction H&S

    □ direct subcontractors such as, for example, tender price, competencies, resources, financial provision for construction H&S and contractual relationships
- management of information such as, for example, inclusion of H&S information in contract documentation, such as site plans, floor plans, details, elevations and schedules
- designer H&S risk management such as, for example, inadequate process and the implications and/or the impact thereof on H&S.

The following are briefly discussed relative to the direct category:
- local communities in terms of project-related work expectations, which have general and H&S skills implications
- socioeconomic issues in the neighbourhood such as, for example, crime and vandalism
- access to the site from public thoroughfares such as, for example, traffic volumes, and peak traffic periods
- health risks arising from client's activities such as, for example, sewage treatment works and nuclear power plants
- constraints in the case of brownfield projects such as, for example, shoppers in shopping malls, hot work in refineries, storage of hazardous materials and fire safety during maintenance
- topography such as, for example, sloping sites, potential rockfalls and flood plains
- ground conditions such as, for example, collapsible soils, contaminated ground, height of water table and previous fill
- existing services such as, for example, position of electricity, fire main, gas, internet, sewage and water supply
- fauna and related issues such as, for example, carnivores/predators, crocodiles, malaria, scorpions and snakes
- viruses and pandemics such as, for example, Covid-19, malaria and sleeping sickness
- availability of services such as, for example, electricity, internet, sewage and water, and the reliability thereof
- heritage issues such as traditional land and the encountering of archaeological findings, fossils and graves
- weather such as, for example, lightning, mist, precipitation (including hail), temperature, humidity, wind chill factor and wind (including prevailing direction)
- natural disasters such as, for example, earth tremors, mud slides, flooding and tornadoes
- unexploded ordinance such as bombs and landmines
- removal and disposal of waste

- adjoining buildings and structures, including:
  - the potential impact of basement construction on, or excavations adjacent to buildings, adjoining properties, including geotechnical interventions such as ground anchors, and soil nailing
  - implications of the height and proximity of adjoining buildings in terms of the use of tower cranes and mobile cranes
- existing structures, such as, for example, degree of stability and fragile components or materials such as roof lights or asbestos-containing materials.

6. Assess the Risk

Risk depends on two elements:
1. The likely severity of harm or ill health/the implication (consequence/impact).
2. The likelihood that the harm or ill health/implication will occur (probability).

A 3 x 3 approach (three-point scale) is a relatively quick approach to risk quantification. Given the nature of BRAs and the scope thereof, the categories of severity have been expanded beyond the traditional H&S-related categories.

In the 3 x 3 approach, likelihood is categorised as 1 (low), 2 (medium) and 3 (high):
- 1 (low): very seldom or never occurs – quarterly or less frequently
- 2 (medium): reasonably likely to occur – weekly to monthly
- 3 (high): often occurs – constantly to daily.

Likely consequence or severity is categorised as 1 (low), 2 (medium) and 3 (high):
- 1 (low): first aid or medical aid injury or illness – minimal or no impact
- 2 (medium): injury or illness causing short-term disability – moderate impact
- 3 (high): fatality, major injury or illness causing long-term disability – major or catastrophic.

Table 5.1 presents a colour coded risk rating based upon the 3 x 3 risk assessment, while Table 5.2 presents the risk level and the form of response per level.

Table 5.1: 3 x 3 risk rating

Consequence/ severity score	Likelihood score		
	1 (low)	2 (medium)	3 (high)
3 (high)	3	6	9
2 (medium)	2	4	6
1 (low)	1	2	3

Table 5.2: 3 x 3 risk level and the form of response per level

Risk level	Form of response
9 (extreme)	Eliminate the hazard/unique interventions are required
6 (high)	Substitute the hazard/additional or special interventions are required
3 – 4 (medium)	Prescribe engineering controls and follow standard work procedures/continuous review of the issue
1 – 2 (low)	Limited response required, such as personal protective equipment/normal interventions

The 5 x 5 approach will not be presented in this chapter, it will be referred to in a later chapter.

Table 5.3 presents a retrospective hypothetical part BRA relative to the Table Mountain Aerial Cableway project, which is informed by, among others, Figure 5.2 in section 2 of this chapter. The hazards included are primarily direct project-wide hazards. The 'owner action' column indicates which stakeholders need to collectively address the hazards. Residual risk has not been included because of the need for brevity and, furthermore, most hazards require interventions on the part of more than one stakeholder, hence the 'Record hazard in' one or more of the 'Designer H&S specification', 'Designer report' and 'Contractor H&S specification' columns.

Table 5.4 presents a hypothetical general BRA focusing primarily on aspects/hazards/issues that affect H&S indirectly. As in the case of Table 5.3, residual risk has not been included because of the need for brevity and, furthermore, most hazards require interventions on the part of more than one stakeholder. However, the 'Record hazard in' one or more of the 'Designer H&S specification', 'Designer report' and 'Contractor H&S specification' columns indicates cases where the hazards need to be noted by and responded to by specific stakeholders and, in some cases, the client included.

Table 5.3: Part post-Table Mountain Aerial Cableway project BRA

BRA Reference	Category	Hazard	Source of hazard	Population impacted	Raw risk classification			Owner action	Record hazard in		
					Likelihood	Consequence	Risk rating		Designer H&S specification	Designer report	Contractor H&S specification
T1	Topography	Sheer rock faces (falls)	Nature	Contractors	3	3	9	Client/CHSA/CPM/Designers/PC/Contractors	●	●	●
T2	Topography	Undulating terrain (slips & trips)	Nature	Contractors	3	1	3	Client/CHSA/PC/Contractors			●
W1	Weather	Mist (loss of visibility)	Nature	Contractors	2	1	2	Client/CHSA/PC/Contractors	●	●	●
W2	Weather	Precipitation	Nature	Contractors	2	2	4	Client/CHSA/CPM/Designers/PC/Contractors	●	●	●
W3	Weather	High temperature	Nature	Contractors	2	1	2	Client/CHSA/PC/Contractors			●
W4	Weather	Low temperature	Nature	Contractors	3	1	3	Client/CHSA/PC/Contractors			●
W5	Weather	Wind	Nature	Contractors	2	2	4	Client/CHSA/CPM/Designers/PC/Contractors	●	●	●
W6	Weather	Unpredictable weather	Nature	Contractors	2	1	3	Client/CHSA/CPM/Designers/PC/Contractors	●	●	●
FA1	Fauna	Caracal	Hazard	Contractors	1	2	2	Client/CHSA/PC/Contractors			●
FA2	Fauna	Litter	Contractors	Fauna	1	1	1	Client/CHSA/PC/Contractors			●
FA3	Fauna	Porcupines	Hazard	Contractors	1	2	2	Client/CHSA/PC/Contractors			●
FA4	Fauna	Scorpions	Hazard	Contractors	2	1	2	Client/CHSA/PC/Contractors			●
FA5	Fauna	Snakes	Hazard	Contractors	2	2	4	Client/CHSA/PC/Contractors			●
FA6	Fauna	Tortoises	Contractors	Tortoises	2	3	6				●
FL1	Flora	Contamination of fynbos and soil, eg by concrete	Hazards	Flora	2	2	4	Client/CHSA/CPM/Designers/PC/PC/PC/Contractors	●	●	●
FA7/FL2	Fauna and flora	Fires	Contractors	Fauna and flora	2	3	6	Client/CHSA/CPM/Contractors			●
A1	Access by the work force to the upper station	Unpredictable weather	Nature	Contractors	2	2	4	Client/CHSA/CPM/PC/Contractors			●
S1	Schedule	Unpredictable weather	Nature	Client Contractors	2	2	4	Client/CHSA/CPM/Designers/PC/Contractors	●	●	●

Table 5.4: General BRA focusing on primarily aspects/hazards/issues that affect H&S indirectly

BRA Reference	Category	Aspect/hazard/issue	Source of aspect/hazard/issue	Population impacted	Raw risk classification			Owner action	Record aspect/hazard/issue in		
					Likelihood	Consequence	Risk rating		Designer H&S specification	Designer report	Contractor H&S specification
S-E1	Socioeconomic	Violence	'Construction mafia'	Project stakeholders	3	3	9	Client/CHSA/CPM/PC			●
S-E2	Socio-Economic	Violence	SMMEs	Project stakeholders	3	2	6	Client/CHSA/CPM/PC			●
S-E3	Socio-Economic	Robbery	Criminals	Project stakeholders	1	3	3	Client/CHSA/CPM/PC/Contractors			●
S-E4	Socio-Economic	Vandalism	Vandals	Project stakeholders	1	2	2	Client/CPM/PC/Contractors			●
S-E5	Socio-Economic	No construction competencies	Local community	Project stakeholders	2	2	4	Client/CHSA/CPM/PC	●	●	●
P1	Procurement	Inadequate time	Project duration	Project stakeholders	2	2	4	Client/CHSA/CPM	●	●	●
P2	Procurement	Fragmentation of contributions	Traditional construction procurement system	Project stakeholders	1	2	2	Client/CPM	●	●	●
P3	Procurement	Inadequate reference to H&S	Conditions of contract/BoQ	Project stakeholders	1	2	2	Client/CHSA/CPM/QS	●	●	
P4	Procurement	Inadequate H&S competencies	CPM	Other project stakeholders	1	2	2	Client	—	—	—
P5	Procurement	Inadequate H&S competencies	CHSA	Other project stakeholders	1	2	2	Client/CPM	—	—	—
P6	Procurement	Inadequate H&S competencies	Designers	PC and contractors	2	2	4	Client/CHSA/CPM	●	●	
P7	Procurement	Poor performance	PC	Other project stakeholders	1	2	2	Client/CPM			●
P8	Procurement	Poor performance	Contractors	Other project stakeholders	1	2	2	Client/CPM/PC			●

7. Review Questions

- Was the inclusion of a BRA in the Construction Regulations 2014 justified? A philosophical and pragmatic response is required.
- What is the intention of the BRA?
- When should the BRA be conducted?

8. Review Exercises

- Conduct a BRA focusing on primarily hazards that affect H&S directly.
- Conduct a BRA focusing on primarily aspects/hazards/issues that affect H&S indirectly.

Bibliography

Bennett, L. 2006. *Risk assessments: Guide to understanding the basics*. Clubview: Safety First Association.

Department of Labour (DoL). 2002. *Section 32 investigation report into the Injaka bridge collapse of 6 July 1998*. Pretoria: DoL.

FlightGlobal. 1999. Four crew die in S African crash. *FlightGlobal*, 17 February 1999, https://www.flightglobal.com/four-crew-die-in-s-african-crash/25108.article#.

Haslam, RA, Hide, SA, Gibb, AGF, Gyi, DE, Pavitt, T, Atkinson, S & Duff, AR. 2005. Contributing factors in construction accidents. *Applied Ergonomics*, 36 (4):401–415.

HM Treasury. 2004. *The orange book: Management of risk – principles and concepts*. Norwich: HM Treasury.

IOL. 2001. 21 saved after roof caves in on Pretoria mall. *IOL,* 20 December 2001, https://www.iol.co.za/travel/south-africa/21-saved-after-roof-caves-in-on-pretoria-mall-79038.

Ismael, Z. 2015. 6 die in FS lightning strike. *iafrica.com,* 19 March 2015.

Singh, S. 2004. *The handbook of competency mapping*. New Delhi: Response Books.

Table Mountain Aerial Cableway. 2014. #ThrowbackThursday – introducing the latest cable cars. *Table Mountain Aerial Cableway,* http://www.tablemountain.net/blog/entry/throwbackthursday-introducing-the-latest-cable-cars.

CHAPTER 6

Designer Health and Safety Specification

John J Smallwood

1. Introduction

In terms of Construction Regulations, 2014 Regulation 5, 'Duties of client':

5(1)(b)   clients are required to prepare a health and safety (H&S) specification based on the baseline risk assessment (BRA)

5(1)(c)   the client is required to provide the designer with the H&S specification

5(1)(d)   the client is required to ensure that the designer takes the H&S specification into account during design

5(1)(f)   the client is required to include the H&S specification in the tender documents.

The Construction Regulations, 2014 define an **H&S specification** to mean 'a site, activity or project specific document prepared by the client pertaining to all health and safety requirements related to construction work'. Given that designers are required to provide clients with a report following the receipt of the H&S specification and the required contents of the report, the H&S specification provided to the designers should be referenced 'designer', while the H&S specification must be included in the tender documents and this version should be referenced 'contractor'. Although the Construction Regulations do not differentiate between the obvious two versions, it is notable that the scope of work for CHSAs, found in the Scope of Work for Categories of Registration of the Project and Construction Management Professions, 2019 (Scope of Work), records: '3.7 Identify and implement precautions necessary for construction project health and safety control and update the construction project tender health and safety specifications' relative to Stage 3, design development, and: '4.3 Finalise construction project tender health and safety specifications and integrate with procurement documentation' relative to Stage 4, tender documentation and procurement.

The H&S specification is a further link in the sequence: BRA → designer H&S specification → designer report → contractor H&S specification → H&S plan → H&S file.

2. Intention/Rationale

H&S specifications are not intended to be a regurgitation of the Construction Regulations but should serve as a guide in terms of the hazards and risks and the client's H&S requirements the PC and contractors need to address relative to a project, namely they are project specific. This is reinforced by the requirement that clients must base the H&S specification on the BRA, which should provide information with respect to the indirect and direct project-wide hazards and risks, and, in some cases, hazards related to specific activities on brownfield projects.

The designer H&S specification is an integral part of the sequence: BRA → designer H&S specification → design HIRA → designer report → contractor H&S specification → H&S plan → H&S file.

3. Evolution

Although the Scope of Work for CHSAs records: '2.6 Prepare draft construction project baseline risk assessment' and '2.11 Prepare the draft construction project health and safety specification' relative to Stage 2, concept and feasibility, the conducting of the BRA and the development of the H&S specification should be initiated during Stage 1, project initiation and briefing, and preferably finalised at the end of Stage 1. This contention is supported by, first, the realities of projects in the form of indirect and direct project-wide hazards and risks, as opposed to solely H&S hazards and risks; and secondly, that the concept design is influenced and informed by the BRA and designer H&S specification. An example is the upper station of the Table Mountain Aerial Cableway upgrade project undertaken in 1997.

4. Contents

The client provided designer H&S specification should address the following:
- project details
- client's considerations and management requirements
- environmental restrictions and existing on-site hazards and risks
- H&S file.

Environmental issues should be addressed, which, although they should be addressed in the environmental impact assessment, designers should obviously flag because they may well be related to H&S, for example fires and removal and disposal of asbestos-containing materials (ACMs).

4.1 Project details

Project details include the following information:

- project location, including:
  - access to the project location such as, for example, infrastructure such as airports, harbours, railway routes above and below ground, rivers and roads; and including conditions, load restrictions and restrictions with respect to transportation of specific goods, which may impact on the supply of materials and plant and equipment to the project such as, for example, wind turbine masts and blades
  - adjacent public thoroughfares such as roads and above ground railway routes in terms of considerations and/or restrictions such as, for example, in the case of façade retention relative to a redevelopment project requiring demolition
  - access to the site in the form of entrances and/or exits from public thoroughfares in terms of restrictions that may affect aspects of design such as, for example, tight turning circles, traffic volumes and peak traffic periods, which may affect, among others, the length of delivery vehicles that can enter and exit the site
  - consideration of the space required for site accommodation, plant and equipment, lay down areas, the delivery of materials to the site and the blocking of public thoroughfares while offloading
  - fauna, flora and related issues such as, for example, protected species in terms of flora in the case of pouring concrete during the Table Mountain Aerial Cableway project that posed a hazard to the indigenous flora, albeit an environmental issue, because the area is part of the Table Mountain National Park
  - existing services such as, for example, electricity, fire main, gas, sewage, stormwater and water
  - socioeconomic issues such as the 'construction mafia', SMMEs, crime and vandalism
  - weather such as, for example, lightning, mist and precipitation (including hail), temperature (including humidity and wind chill factor) and wind (including prevailing direction)
  - other such as, for example, the likelihood of landmines and unexploded ordinance such as bombs. This may be more applicable to countries such as China, in the case of Hong Kong, Germany and the United Kingdom as a result of the Second World War, and this should be remembered within the context of the wars in Southern Africa
- project description in terms of the nature, scope and complexity, such as, for example, new build in the form of a hotel, refurbishment of a hotel or refurbishment of a shopping mall, including alterations and additions. Characteristics include

erf size, current floor area, additional floor area, storeys and capacity, for example, 300 bedrooms in the case of a hotel
- phase and programme requirements such as, for example, separate phases for the basement and the structure, and continued use of a shopping mall during refurbishment, including tenants changing stores
- extent and location of existing records and plans, which will provide information relative to the position of services, existence of post-tensioned slabs and existence of ACMs.

Figure 6.1 indicates the challenges that designers and contractors could face relative to a difficult project.

Figure 6.1: Sheer face construction, upper station, Table Mountain Aerial Cableway upgrade project (Deacon, 1997)

4.2 Client's considerations and management requirements

Client's considerations and management requirements include:
- provision of designers' designing for construction H&S policy, philosophy, competencies and resources, and the design HIRA process, including the documentation thereof, which, theoretically, should have been assessed prior to appointment

- digitalisation in the form of using building information modelling to facilitate designing for construction H&S and the design HIRA
- participation in a multi-stakeholder partnering process, which includes H&S
- community participation in the project, which has implications, in general, in terms of:
  - the skills required relative to the design, details and specification
  - H&S awareness, knowledge and skills required because of, among others, working in excavations or at heights
- project H&S goals, which are likely to include reducing incidents and accidents, and improving H&S practices on projects, a key project objective in terms of H&S, is therefore likely to be a fatality-, injury- and disease-free project, that is, zero harm, and therefore, how designers need to indicate how they will contribute thereto
- H&S monitoring and review: clients may require the monitoring of design-originated hazards and risks, therefore, designers need to indicate the process and documentation thereof
- permit and authorisation requirements as a result of the client's on-site activities on a brownfield project such as, for example, avoidance of on-site welding and other hot work
- fire prevention and protection because of the client's and/or users' activities during occupancy and, therefore, these preventative and protective measures may be relevant in the case of both brownfield and greenfield projects, namely non-interruption of the fire main and/or sprinkler installation
- emergency procedures and egress, in the case of brownfield projects, must not be compromised such as, for example, alterations and additions or refurbishment
- site rules and other restrictions on contractors, suppliers and others such as, for example, access arrangements to those parts of the site that continue to be used by the client, shift work, night work, restricted hours, limiting of in-situ and on site work, dust and noise
- activities on or adjacent to the site during the works such as, for example, construction work relative to the repair or replacement of live sewers
- arrangements for liaison between parties: this requirement could relate to the routing and approval of specialist contractors' detailed shop drawings such as structural steelwork, air conditioning and electrical, in which cases, the CPM or principal agent, PC, specialist contractors and respective engineers are pivotal. A further category could relate to temporary works in relation to the permanent works such as the attachment of tower crane stays to the permanent structure, in which case, the CPM or principal agent, PC and structural designer are pivotal
- submission of a project H&S closeout report, or contribution thereto.

4.3 Environmental restrictions and existing on-site hazards and risks

Environmental restrictions and existing on-site hazards and risks should address the health hazards and safety hazards.

4.3.1 Health hazards

Examples of health hazards that may be encountered include:

- asbestos, in the case of brownfield projects, especially those entailing alterations and additions, or refurbishment
- existing storage of hazardous materials such as, for example, fuel tanks in a tank farm
- contaminated land in the case of both brownfield and greenfield projects, which is likely to affect activities
- health risks arising from the client's activities such as, for example, sewage treatment works and nuclear power plants.

4.3.2 Safety hazards

Examples of safety hazards that may be encountered include:

- adjacent land uses may include existing buildings or structures, therefore, the foundations thereof, and the related services, should not be compromised by the proposed project's earthworks and, furthermore, a possible structural failure courtesy of such neighbouring buildings or structures may be experienced during the project's earthworks
- location of existing services such as electricity, fire main, gas, internet, sewage, stormwater and water, which may be underground, above ground or overhead, the latter being visibly obvious. The H&S file should be provided in the case of a recent brownfield project. If such an H&S file does not exist, existing site plans and plans of services and utilities should be provided
- ground conditions, which the client may be aware of and which include some or all of the following:
  - the soil type, which has stability implications
  - ground water level
  - possible ingress of water
  - old mine workings
  - related possible contamination, in the case of harbouring of leaking fuel tanks or other hazardous materials
  - underground obstructions, such as foundations
  - possible subsidence
- existing structures such as, for example, the degree of stability, in the case of elements that need to be retained and included in the proposed structure, such as

a historical façade, or parts of a structure that will be demolished. Other elements include:

- fragile elements, components or materials, such as existing suspended timber floors, including the supporting structure,
- roof lights.

4.4 The H&S file

The format of the H&S file should be indicated, whether electronic or print, and thereafter, the structure should be indicated. The volume of information may require a document that summarises the key elements, in the form of a quick guide. Furthermore, H&S files could be presented in two parts, the first for information that may need to be accessed on a day-to-day basis, such as, for example, operational and maintenance manuals and the second long-term use such as, for example, documentation required when alterations and/or additions and/or refurbishment work is contemplated.

Thereafter, the required H&S file contents should be recorded, among others:

- as-built drawings and plans
- design criteria such as, for example, design loadings and calculations
- construction methods and materials used such as, for example, post-tensioned reinforced concrete slabs
- potential hazards included in the structure such as, for example, post-tensioned cables
- record of hazardous processes such as, for example, removal or encapsulation of ACMs
- equipment and maintenance facilities
- maintenance procedures and requirements
- manuals (operating and maintenance) for plant and equipment
- location and nature of utilities and services, including an as-laid site plan.

5. Review Questions

- What is the purpose of the designer H&S specification?
- What are the pre-requisites to realise an informed designer H&S specification?
- Based upon your experience in general, are designer H&S specifications informative?

6. Review Exercises

- Compile a designer H&S specification for a project based on your BRA.
- Conduct a review of a designer H&S specification.

Design Hazard Identification and Risk Assessment

John J Smallwood

1. Introduction

In terms of the Construction Regulations, 2014 Regulation 6, 'Duties of designer', 6(1) designers of a structure must:

(a) ensure that the health and safety (H&S) standards incorporated into the regulations are complied with in the design

(b) take the H&S specification into consideration

(c) include in a report to the client before tender stage:
   i.   all relevant H&S information about the design that may affect the pricing of the work
   ii. the geotechnical-science aspects
   iii. the loading that the structure is designed to withstand

(d) inform the client of any known or anticipated dangers or hazards relating to the construction work and make available all relevant information required for the safe execution of the work upon being designed or when the design is changed, which may require design and construction method statements

(e) modify the design or make use of substitute materials where the design necessitates the use of dangerous procedures or materials hazardous to H&S

(f) consider hazards relating to subsequent maintenance of the structure and make provision in the design for that work to be performed to minimise the risk

(g) (j) during the design stage, take cognisance of ergonomic design principles in order to minimise ergonomic-related hazards in all phases of the life cycle of a structure.

To meet these requirements requires clients, construction health and safety agents (CHSAs), construction project managers (CPMs), designers and quantity surveyors (QSs) to:
- identify hazards and assess the risk
- eliminate or mitigate the hazards and risks by adopting the hierarchy of control

- record the residual risk, if any, that should be recorded in the designer report and the contractor H&S specification
- document the design hazard identification and risk assessment (HIRA) processes.

The Construction Regulations and the subsequent guidelines published on 2 June 2017, do not address the duties of designers relative to the six recognised project stages in South Africa, which may lead to the perception that the duties do not apply during the first two stages. However, design HIRA should be addressed during all six project stages, namely:

1. project initiation and briefing
2. concept and feasibility
3. design development
4. tender documentation and procurement
5. construction documentation and management
6. project closeout.

Furthermore, a design HIRA should be required following any redesign during the construction documentation and management stage.

Ergonomic related hazards require analysis and evaluation, and must be addressed during HIRAs. Therefore, a design HIRA is a further link in the sequence: BRA $\rightarrow$ designer H&S specification $\rightarrow$ design HIRA $\rightarrow$ designer report $\rightarrow$ contractor H&S specification $\rightarrow$ H&S plan $\rightarrow$ H&S file.

2. The Role of the Occupational Health and Safety Inspectorate of the Department of Employment and Labour

To date there have been no reports of visits to design practices or assessment of designers' construction H&S-related interventions by the Occupational Health and Safety (OH&S) Inspectorate of the Department of Employment and Labour (DEL) in South Africa. This lack of intervention is unlikely to have contributed to engendering designing for construction H&S, including design HIRAs.

Perhaps a lesson can be learnt from the Construction Division of the Health and Safety Executive (HSE) in the United Kingdom (UK) which conducted 122 visits to sites between 19 and 30 April 2004 to interview 116 representatives of design practices in the north of the UK. These visits are indicative of the inclusive approach in the UK with respect to ensuring that all parties in the construction H&S effort are involved in improving the overall construction H&S performance of the industry. Evidently, the same cannot be said for the DEL.

3. Designers' Contribution to Risk Control

Risk control measures should be collective rather than personal, and advocate the following hierarchy of risk control:

- changes that eliminate the hazard
- substitution of a less hazardous design feature
- engineering controls in the form of isolation, barriers, guarding or segregation, which basically separate people from the hazard
- reduced exposure, ie changes that reduce the period of exposure to a hazard, or the number of people exposed thereto
- safe working procedures, including appropriate training and competent close supervision
- written procedures, which could include design and construction method statements, the provision of information, instructions, warnings, signs or labels
- use of personal protective equipment (PPE).

4. Design Philosophy Statements

Design philosophy statements are advocated because they provide insights to the approach adopted by the designer(s), can form part of the design HIRA-related documentation and can be included in the H&S file. Furthermore, in the case of South Africa, they can be included in the designer report and the contractor H&S specification. Potential inclusions include, among others:

- considerations with respect to the contractor's site layout, site establishment, storage and laydown areas required, the latter being dependent on the design length of, among others, reinforcing steel bars and trusses
- accommodation of major permanent plant and equipment as a result of implications related to the hoisting thereof, if positioned at the top of a building or structure
- specification of driven piles, as opposed to augered piles, in the case of contaminated ground
- considerations with respect to limiting the depth of excavations
- considerations with respect to promoting constructability of the structural frame
- optimisation of off-site prefabrication to minimise work at height
- considerations with respect to the enclosing fabric, the attachment thereof to the structure and the cleaning thereof during the use phase
- minimisation of hot work, such as welding on site
- the design of common ducts for services to limit the number of openings in floor slabs

- ergonomic considerations relative to ducts, walling and ceilings accommodating single or multiple services on installation of services during construction and their maintenance during the use phase
- use of drywalling, as opposed to masonry partition walling, to:
  - reduce chasing for services
  - enhance constructability, productivity, quality- and time-related performance
  - reduce ergonomic hazards
  - reduce the volume of solid waste
- specification of pre-glazed door and window units to reduce on-site glazing and glazing at heights
- specification of self-cleaning glazing
- specification of finishes that require limited, if any, further finish, for example, self-finished drop in suspended ceiling tiles
- consideration of access to light fittings during maintenance
- specification of light emitting diode (LED) light fittings where possible, to mitigate frequent replacement of bulbs
- frequency of maintenance of flora specified relative to landscaping.

5. Design HIRA Practices

The following constitute the primary design HIRA and related practices, namely:
- document the design HIRA process
- review the client H&S specification
- maintain a register of project hazards and risk
- gather H&S information relative to projects
- assemble H&S expertise, for example consult a CHSA
- identify hazards
- avoid/eliminate hazards
- amend designs
- amend details
- substitute materials
- identify residual hazards
- identify risks from residual hazards
- assess/prioritise/investigate selected risks
- focus on significant/unusual/difficult risks
- prepare design and construction method statements
- provide information on residual risks, for example on drawings
- prepare a designer report (H&S) for clients
- provide H&S information for tender documentation
- maintain registers for construction and project hazards and risks
- review the construction phase H&S plan

- monitor construction activities relative to design HIRAs
- conduct site risk assessments and actions/directions
- revisit the process if the design changes at any point
- consider H&S during maintenance
- contribute to the H&S file
- compile a project H&S lessons learnt report.

5.1 Document the design HIRA process

Designers should be able to substantiate their compliance with the requirements of the Construction Regulations for various reasons. First, to reflect professionalism and best practice; secondly, should the OH&S Inspectorate conduct an overview visit to a design practice to review designers' compliance; and thirdly, should an accident that has its origin, partly or wholly in the design process, occur.

Consequently, a documented trail should be available, which includes the process followed leading to the design HIRA, the design HIRA process itself and the responses. Responses should be included in the designer report, which includes design and construction method statements, annotations to drawings and comments or remarks in schedules.

The non-inclusion of this as a requirement of the Construction Regulations constitutes a shortcoming, which should be addressed.

5.2 Review the client H&S specification

This is a requirement of Construction Regulations, 2014 Regulation 6, 'Duties of designer', 6(1)(b). However, the extent to which the H&S specification assists designers to achieve optimum designing for construction H&S and to comply with the requirements of the Construction Regulations depends on the client's baseline risk assessment (BRA) and the contents of the H&S specification. Client requirements, which may require designer interventions should also be addressed such as, for example, the phased refurbishment of a shopping mall while the mall is in use.

5.3 Assemble H&S expertise

Ideally, a CHSA should be appointed, even if a permit to do construction work is not required. Many clients have H&S or health, safety and the environment departments or units and, in some cases, have contractor H&S programmes and designated H&S practitioners who manage the contractors. Therefore, these clients are likely to able to contribute in terms of gathering H&S information. However, specific expertise may be required to take into consideration the unique features of a proposed building or structure, eg contaminated land, which would require the appointment of an environmental engineer.

5.4   Gather H&S information relative to projects

This process is multifaceted in nature and, in theory, commences with the client's BRA, followed by the designer H&S specification. The client brief is a further opportunity to identify H&S information and issues. Furthermore, if the building or structure is a brownfield project constructed post-July 2003, then, in theory, an informative H&S file should be available, which should provide information regarding potential hazards in the building or structure to be altered, added to or refurbished such as, for example, the position of electrical cables and fire mains, encapsulated asbestos fibre and post-tensioned cables.

5.5   Identify hazards

The identification of hazards and the quantification of risk is the essence of designing for construction H&S.

There are two categories of hazards, namely:

1. hazards that are likely to be recognised by a competent contractor but which might be reduced by designers' interventions
2. hazards that are unlikely to be recognised by a competent contractor.

Furthermore, it is important to:

- identify and eliminate hazards at an early stage
- identify interdisciplinary issues where two or more design elements interface
- consider all the activities entailed in constructing the design, and the hazards and risks that might result from their interaction.

Identification of hazards can be realised by considering categories of hazards as follows:

- falls from height because of activities/trades such as, for example, structural frame erection, roof work, cladding and window cleaning during the use phase
- harmful substances such as, for example, contaminated land, paint application and stone cutting
- manual handling of building blocks, long roof sheets, kerb stones and paving slabs.

An alternative approach is using a work breakdown structure, which entails identifying hazards by referring to elements/finishes/activities relative to work packages.

Work packages include:

- substructure, including the excavations, bases and footings, columns and/or walls: contaminated land; collapsible soils; chromates in cement used in concrete; and handling heavy blocks
- structure, including the frame, walls and roof: ergonomic hazards relative to falsework falls from height and to the same level; chromates in cement used in concrete, mortar and plaster; and handling long roof sheets
- partition walling, door frames and ceilings, including rendering: ergonomic hazards from laying masonry units or erecting drywalling and ceilings; chromates in cement, mortar and plaster; and applying plaster and paint; and falls to same level
- envelope, including the walls and/or cladding and roofs: ergonomic hazards from laying masonry units and applying plaster and paint; falls from height; and noise and dust from cutting natural stone
- services, such as air conditioning, electrical, plumbing and drainage: ergonomic hazards; falls from height; and falls to same level
- finishes and fittings, including kitchen cupboards, built in cupboards, doors and light fittings: ergonomic hazards; and falls to same level
- site works, including site clearance, installation of services and stormwater drainage, kerbing, road surfacing and landscaping: collapsible soils; handling heavy kerbs; exposure to vapours and aerosols; cutting paving blocks; and handling thorny and/or heavy palms.

5.6 Assess/prioritise/investigate the risk

Although it can be argued that hazards and the related risks should be progressively addressed, prioritisation thereof through quantification in the form of a risk rating can provide an agenda for addressing the major hazards and risks on a project.

A 5 x 5 approach (5-point scale) is a logical approach since it is more detailed than a 3 x 3 approach discussed in an earlier chapter.

Likelihood:

- 1 (rare): very unlikely, if ever – six-monthly or less frequently
- 2 (seldom): unlikely – monthly
- 3 (occasionally): likely – weekly
- 4 (frequently): very likely – daily
- 5 (continuously): almost certain – constantly or hourly.

Likely severity:

- 1 (minor): first aid treatment
- 2 (moderate): medical aid treatment
- 3 (severe): temporary disablement
- 4 (very severe): permanent disablement
- 5 (extreme): fatal.

Table 7.1 presents a colour coded risk rating based upon the 5 x 5 risk assessment, while Table 7.2 presents the risk level and the form of response per level.

Table 7.1: 5 x 5 risk rating

Consequence score	Likelihood score				
	1 (rare)	2 (seldom)	3 (occasionally)	4 (frequently)	5 (continuously)
5 (extreme)	5	10	15	20	25
4 (very severe)	4	8	12	16	20
3 (severe)	3	6	9	12	15
2 (moderate)	2	4	6	8	10
1 (minor)	1	2	3	4	5

Table 7.2: 5 x 5 risk level and response per level

Risk level	Response per level
20 – 25 (extreme)	Eliminate the hazard
12 – 16 (high)	Substitute the hazard
5 – 10 (medium)	Prescribe engineering controls
1 – 4 (low)	Limited response required, eg PPE

5.7 Focus on significant/unusual/difficult risks

The risks can be prioritised in terms of the risk ratings and addressed progressively. However, red risk level should be addressed immediately, which is 9, in the case of the 3 x 3 approach, and 20-25, in the case of the 5 x 5 approach.

5.8 Avoid/eliminate hazards

Avoiding or eliminating hazards is the first step in the hierarchy of control, followed by substitution. Therefore, should the risk rating be 9, in the case of the 3 x 3 approach, and 20-25, in the case of the 5 x 5 approach, the source of the hazard should be eliminated.

However, there are materials or components, or similar, for which it is not possible to reduce the raw risk through elimination or substitution, or even engineering controls. An example being palm trees. The challenge with respect to quantifying the weight of a palm is that the calculation is dependent upon the height, girth and size of the root ball; the soil or rock contained within the root ball; and the number of fronds. However, palms are a common landscaping plant, and are often specified for both landscaping and plant-scaping, which, in the case of the latter, has an evolving dead load implication from a permanent works design perspective.

Figure 7.1 shows a cotton palm in the process of being offloaded by a truck-mounted crane.

Figure 7.1: Large date palms alongside a walkway, Gqeberha (Smallwood, 2005)

5.9 Amend designs

Suspended reinforced concrete slabs incorporating L- and T-beams require more production hours and effort, and exposure to ergonomic hazards such as bending and twisting, adopting uncomfortable positions and repetitive movements on the part of both form workers and steel fixers. Amendment of the design by opting for a flat slab will simplify the support work and formwork, facilitate the use of table forms or single level falsework and avoid the need for fixing traditional steel reinforcing by using post-tensioning. Such an amendment will reduce the amount of labour hours and exposure to hazards, activity time and cost, and promote constructability.

A further example is that of in situ reinforced concrete stair flights and landings, which entail challenging raking falsework and formwork, and non-ergonomic activities. Amending the design from in situ to precast stair flights will eliminate the challenges, ergonomic hazards and the amount of time working at height. Precasting at ground level is likely to reduce costs and enhance productivity, quality and schedule.

5.10 Amend details

Figure 7.2 shows a patterned eaves lining requiring extensive cutting of fibre cement board, sanding of the strip edges, unnecessary reaching overhead and repetitive movements to fix the eaves lining to the timber framework. These hazards can be mitigated, to a major extent, by detailing a flat soffit for the eaves lining.

5.11 Substitute materials

Substitution of materials is the second step in the hierarchy of controls, following elimination. Large dimensioned concrete blocks may have been envisaged for a loadbearing masonry structure, however, because of their mass, they may constitute a hazard. In such a case, clay maxi bricks could be a potential substitute.

Figure 7.2: Patterned eaves lining, Gqeberha (Smallwood, 1992)

5.12 Identify residual hazards and risk

A residual hazard is a hazard that remains after the raw hazard has been substituted, or after the development of engineering controls. The risk should be quantified and the residual hazard should be flagged in the design HIRA documentation, on drawings, in schedules, in the designer report and in the contractor H&S specification.

An example of a raw hazard and risk is the forming of an opening in a wall of an existing reinforced concrete lift shaft to install a door opening in a different direction. Forming such an opening manually using paving breakers would expose the workers to hand-arm vibration (HAV), noise and dust. Furthermore, the structural integrity of the lift shaft may be compromised as a result of the process, in addition to being time consuming. Substituting the manual method with an alternative method or, it could be argued, introducing an engineering control in the form of a thermal lance as depicted in Figure 7.3, mitigates the raw hazard and risk. A thermal lance is a long steel tube filled with a series of metal elements connected to an oxygen source in the form of a cylinder, including a regulator. The regulator enables the operator to regulate the flow of oxygen and, therefore, the rate of consumption of the lance. To ignite the thermal lance, the end of the tube is lit with a high temperature source, such as an oxyacetylene torch. The iron in the steel burns in the oxygen coming down the pipe to produce enormous heat, and a liquid slag of

Figure 7.3: Thermal lance in use to form an opening in a wall of a reinforced concrete lift shaft, Gqeberha (Smallwood, 1987)

iron oxides and other materials, which dribbles and splashes out. The temperature reached at the centre of the combustion zone is approximately 4 000 °C, enabling the lance to cut through the reinforced concrete. It should be noted that concrete melts at 1 800 to 2 500 °C, while steel melts at less than 1 500 °C. The lance can penetrate 300 mm thick concrete in 30 seconds (Thermo-Bore (Pty) Ltd, 1987). However, using the thermal lance results in residual hazards and risks in the form of heat and molten metal, requiring the wearing of appropriate, extreme heat-resistant PPE. The raw and residual hazard and risk is indicated in reference S1 in Table 7.3.

A further example of a raw hazard and risk is a standard precast concrete kerb, which constitutes a hazard when handling it because of its mass, as indicated in reference SW1 in Table 7.3.

Substituting such a kerb with an extruded in situ kerb will avoid the need for hazardous manual handling, and this will also result in a reduction in ergonomic hazards, such as bending and a degree of twisting, which constitute a residual hazard and risk.

Table 7.3: Output of the design HIRA process

Reference	Activity	Hazard	Population	Raw risk classification			Design response to mitigate the risk rating	Design action owner	Residual risk classification			Future action
				Likelihood	Consequence	Risk rating			Likelihood	Consequence	Risk rating	
S1	Create opening in a reinforced concrete lift shaft wall	HAV, whole-body vibration and dust	Workers	3	2	6	Prescribe the use of a thermal lance	Structural engineer	1	2	2	Include in designer report Include in H&S specification Address in principal contractor (PC) H&S plan
M1	Concrete block walling	Handling heavy blocks	Workers	3	2	6	Substitute with maxi bricks	Architect	3	1	3	Designer to monitor during construction
F1	Laying vinyl floor sheeting	Exposure to irritants and toxins	Workers	3	2	6	Specify water-based adhesives as opposed to solvent-based adhesives	Architect	3	1	3	Designer to monitor during construction
SW1	Concrete kerbing	Handling heavy kerb sections	Workers	3	2	6	Substitute with extruded in situ kerbs	Civil engineer	3	1	3	Designer to monitor during construction
SW2	Concrete block paving	Cutting paving blocks	Workers	3	2	6	Amend pattern	Civil engineer	3	1	3	Designer to monitor during construction
SW3	Asphalt road surfacing (hot mix)	Exposure to vapours and aerosols	Workers	3	2	6	Warm mix not accepted (none)	Civil engineer	-	-	-	Include in designer report Include in H&S specification Address in PC H&S plan
L1	Planting cotton palms	Weight Striking or crushing of workers	Workers	2	3	6	Cannot eliminate or substitute (none)	Landscape architect	-	-	-	Include in designer report Include in H&S specification Address in PC H&S plan

5.13 Prepare design and construction method statements

Although designers may argue that it is not their function to prescribe to contractors how they should execute the construction of the permanent works design, the design may be such that specific interventions are required to ensure the structural integrity of the permanent structure and to prevent, for example, a possible collapse during construction.

An example is that of a composite slab, as shown in Figure 7.4, consisting of precast concrete ribs supporting hollow core concrete blocks, in situ concrete between the blocks and an in situ concrete overlay of the entire area, incorporating steel mesh reinforcing. The following should be included in a composite slab design and construction method statement:

- description of the slab
- loads such as, for example, x kg per square meter (composite system)
- hazards and risks such as, for example, mass of planks (per meter) and blocks (unit)
- interventions and duration such as, for example, temporary support, including bracing, for x days
- precautions such as, for example, x days before y load per square meter after pouring in situ overlay
- general requirements such as, for example, pre-pour inspection by the structural engineer.

Figure 7.4: Precast planks/ribs and blocks to a composite slab, Plettenberg Bay (Hamp-Adams, 1994)

A further example is that of structural steel trusses as part of a structural steel structure, which may require the simultaneous installation of the horizontal bracing between the trusses. This should be clearly indicated by the designer.

5.14 Cross-reference raw and residual risks

Project documentation such as site plans, floor plans, foundation or base layouts, reinforcement layouts, elevations, specifications and schedules are examples of cross referencing raw and residual hazards.

5.15 Provide H&S information for tender documentation

This practice was referred to in the introduction to this chapter and should include geotechnical-science information, such as trial hole results, findings emanating from archival and physical surveys, a summary of the raw and residual hazards and risks in the form of a designer report project hazard and risk register, and the design and construction method statements.

5.16 Maintain registers for construction and project hazards and risks

In general, designers should develop and maintain a register of typical and unique construction hazards and risks. However, a project-specific hazards and risks register should be developed and included in the designer report for all projects, regardless of scale.

Table 7.4 presents typical hazardous activities, examples of the related hazards and examples of responses to such hazards. Such registers constitute a reference library, a reference for designer induction and, a source of awareness, and enable designing for construction H&S knowledge transfer.

Table 7.4: Construction hazardous activities, examples and responses to reduce risks register

Hazardous activity	Example of hazard	Example of response
Excavations	▪ Collapsible soil ▪ Contact with services	▪ Geotechnical survey and limit excavation depth ▪ Archival and physical survey
Underpinning	Collapse of the structure	Design and construction method statement (sequence of construction)
Manual handling	Handling heavy blocks	Specify light blocks or maxi bricks
Manual handling	Handling heavy kerb sections	Specify extruded in situ kerbs
Working with handheld tools	HAV syndrome	Specify surface finishes that do not require scabbling as bush hammered architectural concrete does

$\rightarrow$

Hazardous activity	Example of hazard	Example of response
Working at heights	Falls from heights	Pre-assembly of elements and precast sections
Working with hazardous chemical substances	Exposure to irritants and toxins	Specify water-based adhesives as opposed to solvent-based adhesives
Cutting paving blocks	Dust and noise	Minimise cutting through specification of pattern requiring less cutting, eg basketweave/parquet
Asphalt road surfacing (hot mix)	Exposure to vapours and aerosols	▪ Include in the H&S specification as a warm mix is not accepted ▪ Address in PC H&S plan

The project register should be categorised in terms of activities as per Table 7.5 on the next page. A brief description or descriptions of what the activity entails should be provided, the response to the hazard, the respondent in terms of the response to the hazard, the future action relative to each respondent and the status of the future action.

5.17 Review the construction phase H&S plan

Although not required in terms of the Construction Regulations, which constitutes further shortcomings therein, designers should be involved in the review of the PC's financial provision for H&S at tender stage, in addition to the review of the PC's H&S plan. The rationale therefore is that the financial provision for H&S and the H&S plan constitute a response to the contractor H&S specification, which is informed by the BRA, the designer H&S specification, the design HIRA and the designer report, and, consequently, designers should close the loop, namely $0°$ through to $360°$.

5.18 Monitor construction activities relative to design HIRAs

Designers should make a concerted effort to review the impact of their design, detailing, specifying and scheduling on the physical construction process and its activities by undertaking regular site visits. Doing so will enable them to realise continuous improvements in terms of design HIRA and the submission of a project H&S closeout report, or their contribution thereto.

5.19 Conduct site risk assessments and actions/directions

While monitoring construction activities relative to design HIRAs, designers should conduct pre-activity and activity HIRAs because doing so will afford them the opportunity to review the activity in person, ie compare the design versus the real-time hazards and risk ratings.

Table 7.5: Project hazard and risk register

Activity	Description	Hazard	Response	Respondent	Future action	Status
Site clearance	Clearing bush	Venomous snakes Thorn trees	Include in designer report	▪ Client ▪ PC	▪ Include in H&S specification ▪ Include in H&S plan	▪ Client to action ▪ Response awaited from PC
Earthworks	Excavations	Collapsible soil	Include in designer report	▪ Client ▪ QS ▪ PC	▪ Include in H&S specification ▪ Address in BoQ ▪ Include in H&S plan	▪ Client to action ▪ QS to action ▪ Response awaited from PC
Suspended RC slab construction	Erection of falsework	Handling, erecting & striking of falsework	Flat slabs (enable use of table forms)	None	None	None
	Post-tensioning	Injuries during stressing	Include in designer report	▪ Client ▪ PC	▪ Include in H&S specification ▪ Include in H&S plan	▪ Client to action ▪ Response awaited from PC
Masonry walling	Laying blocks	Manual handling of heavy blocks	Substitute with maxi bricks	None	None	Completed
Floor coverings	Laying vinyl floor sheeting	Exposure to irritants and toxins	Specify water-based adhesives as opposed to solvent-based adhesives	None	None	Completed
Site works	Kerbing	Handling heavy kerb stones	Specify extruded in-situ kerbs	None	None	Completed
	Laying pavers	Cutting pavers due to herringbone pattern	Revise pattern to basketweave/parquet	None	None	Completed
	Asphalt road surfacing (hot mix)	Exposure to vapours and aerosols	Include in designer report (warm mix not accepted)	▪ Client ▪ PC	▪ Include in H&S specification ▪ Include in H&S plan	▪ Client to action ▪ Response awaited from PC
Landscaping	Planting cotton palms	▪ Weight ▪ Striking or crushing of workers	Included in designer report	▪ Client ▪ PC	▪ Include in H&S specification ▪ Include in H&S plan	▪ Client to action ▪ Response awaited from PC

5.20 Revisit the process if the design changes at any point

This is a requirement in terms of the Construction Regulations, as indicated in the introduction to this chapter. Although it may be deemed stringent, it is logical because designs change, which results in variation orders that may result in out-of-sequence work, which invariably has H&S implications. Furthermore, new hazards and risks may arise as a result of different activities.

5.21 Consider H&S during maintenance

This, too, is a requirement in terms of the Construction Regulations, as indicated earlier in this chapter. Given the multidisciplinary nature of construction projects, the scope of consideration is extensive and, so too, the contributions, for example:

- access to piping in vertical service ducts and services in ceiling spaces
- window cleaning
- edge protection to the flat roof of a multistorey building.

5.22 Contribute to the H&S file

The designer H&S specification records the requirement for designers to contribute thereto. The design HIRA provides the foundation for the H&S file, which is intended to be a lifetime guide because it informs the designer report. The contributions are addressed in detail in the designer report.

5.23 Compile a project H&S lessons learnt report

Clients may require the submission of a project H&S closeout report, or contribution thereto. If not, designers should compile a lessons learnt report, which interrogates, among other, the design-originated risks encountered on site.

6. Review Questions

- What is the purpose of a design HIRA?
- What are the prerequisites to realise effective design HIRA?
- What is the purpose of design philosophy statements?

7. Review Exercises

- Conduct a design HIRA for a project.
- Conduct a designing for construction H&S review of a final design.

Bibliography

Charnock, D. 2004. *Designer initiative 2004 report*. Manchester: Health and Safety Executive (Construction Division).

Gilbertson, A (ed), Arup Group Ltd & Construction Industry Research and Information Association. 2005. *CDM regulations - work sector guidance for designers*. 2nd edition. London: Construction Industry Research and Information Association (CIRIA).

Smallwood, JJ. 2016. Design hazard identification and risk assessment (HIRA). In: P Hájek, J Tywoniak, A Lupíšek & K Sojková (eds) *Proceedings Central Europe Towards Sustainable Building 2016 Conference*, Prague, Czech Republic (22–24 June 2016): 903–910.

Thermo-Bore (Pty) Ltd. 1987. Edenvale: Thermo-Bore (Pty) Ltd.

CHAPTER 8

Designer Report

John J Smallwood

1. Introduction

In terms of Regulation 6, 'Duties of designer', of the Construction Regulations, 2014, designers must:

(1)(c) include in a report to the client before tender stage:
- (i) all relevant health and safety (H&S) information about the design that may affect the pricing of the work
- (ii) the geotechnical-science reports and information
- (iii) the loading that the structure is designed to withstand

(1)(d) inform the client in writing of any known or anticipated dangers or hazards relating to the construction work and make available all relevant information required for the safe execution of the work upon being designed or when the design is changed

(1)(e) modify the design or make use of substitute materials where the design necessitates the use of dangerous procedures or materials hazardous to H&S

(1)(f) consider hazards relating to subsequent maintenance of the structure and make provision in the design for that work to be performed to minimise the risk

(1)(j) consider ergonomic design principles to minimise ergonomic-related hazards during all phases of the life cycle of a structure.

It should be noted that life cycle refers to the entire life of a building, including its design and construction.

The Construction Regulations do not include a definition of a designer report and the Guidelines to the Construction Regulations are silent with respect to what constitutes 'relevant H&S information about the design that may affect the pricing of the work'. Furthermore, although (1)(d) refers to 'inform the client in writing' and (1)(e), (f) and (j) do not require notification by, but merely actions on the part of, the client, it would be appropriate and constitute better practice to include the related issues in the report referred to in (1)(c).

2. Intention of a Designer Report

The designer report is intended to constitute a response to the designer H&S specification, which in turn, should be based on the client's baseline risk assessment (BRA). The designer H&S specification, in turn, along with the designer report, should facilitate the development of the contractor H&S specification.

The designer report is an integral part of the sequence: BRA → designer H&S specification → design hazard identification and risk assessment (HIRA) → designer report → contractor H&S specification → H&S plan → H&S file.

3. Contents of a Designer Report

The Construction Regulations provide some insight into the minimum requirements, but not the ideal requirements or what constitutes better practice. Furthermore, it is not stated in the Construction Regulations whether the designer report should be consolidated or individualised per design discipline. The issue is that the outputs of the collective or individualised reports should be included in the contractor H&S specification.

3.1 Designer H&S policy

The designer H&S policy will immediately indicate whether a design practice is informed with respect to designing for construction H&S or not. An example is as follows:

Example 8.1: Designer H&S policy

Given that people are the most important project resource (value) we, as ABD Architects, have a vision of a fatality-, injury- and disease-free project (vision), which we will promote through designing for construction health and safety, including design hazard identification and risk assessment, documentation of processes, digitalisation of the design and related processes, maintenance of a construction hazards library and allocation of adequate resources, including staff training and development. Through continual improvement (mission), we can contribute to improving health and safety practices and compliance on projects, and to reducing incidents and accidents (goals) by reducing design-originated hazards and their impact. This will promote the sustainability of the environment, the industry and the health and well-being of all project stakeholders, including the occupants, users and the public (purpose).

3.2 Design philosophy statement

The design philosophy statement, as discussed in an earlier chapter, should follow the designer H&S policy, which effectively constitutes what was done to mitigate project-wide, indirect and direct hazards and risk, and design-originated hazards and risks.

3.3 The designing for construction H&S process

A summary should be provided in terms of the interventions per project stage, and should address design reviews, design coordination meetings, design HIRA and the documentation process.

3.4 Response to the designer H&S specification and the significant design and construction hazards and risks

The client provided designer H&S specification, which should address the following, further informs:

- project details
- client's considerations and management requirements
- environmental restrictions and existing on-site risks
- H&S file.

Therefore, the designer report should address these items in the contractor H&S specification, namely the significant design and construction hazards and risks in terms of H&S. A further item that needs attention is that of environmental issues, which, although they should be addressed in the environmental impact assessment, designers should obviously flag as they may well be related to H&S, eg fires and the removal and disposal of asbestos-containing materials (ACMs).

3.4.1 Project details

Project details include the following items which designers should confirm having taken cognisance of and/or how they are envisaged to be addressed, namely:

- project location including:
    - access to the project location such as, for example, confirmation of consideration of load restrictions, and abnormal load restrictions with respect to the transportation of elements, materials, components and plant and equipment to the project. Examples include glass fibre swimming pools and wind turbine masts and blades
    - adjacent public thoroughfares such as roads, and above ground railway routes such as, for example, the temporary works included in the permanent works design in terms of the lateral restraint and support and provision for pedestrian

movement in the case of façade retention relative to a redevelopment project entailing demolition, and how the probable hazards and risks related thereto, and to access, have been limited. Figure 8.1 provides a visual example of the façade retention.

Figure 8.1: Façade retention entailing lateral restraint and passage of pedestrians, London (Smallwood, 2019)

□ access to the site from public thoroughfares such as, for example, confirmation of consideration and response in terms of the design length of components, such as steel reinforcing and roof trusses, in the case of tight turning circles, traffic volumes and peak traffic periods that affect, among others, the length of delivery vehicles that can enter and exit the site

- the footprint of the building or structure and a permanent works' designer potential site layout in terms of available space for site accommodation, plant and equipment, lay down areas, delivery of materials to the site and the blocking of public thoroughfares while offloading
- fauna, flora and related issues: The design interventions to limit the impact of the construction process and its activities on the fauna and flora such as, for example, off-site fabrication in general, including precast concrete elements to limit the pouring of in situ concrete
- services and utilities such as, for example, electricity, fire main, gas, sewage, stormwater and water. Identification of the position of any such existing services on the site plan, including the results of site surveys, and an indication of the processes to be followed to avoid disruption thereof during the construction process, especially during earthworks, excavations included. The rerouting of any powerlines or overhead power supply lines should be recorded and details provided to prevent plant and equipment coming into contact therewith
- socioeconomic issues such as the 'construction mafia', SMMEs crime, and vandalism. These can be countered with design interventions such as:
  - concrete block paving as opposed to asphalt surfaced parking areas to accommodate SMME participation
  - construction of the permanent site perimeter at the inception of the project, and the design and fixing of window units to prevent removal after installation to combat crime during the project
  - the design and installation of window screens to prevent vandalism
  - choice of wall finish to mitigate the impact of graffiti
- weather such as, for example, lightning, mist, precipitation (including hail), temperature (including humidity and wind chill factor) and wind. These can be responded to by design interventions such as limiting of in situ work in high rainfall areas, external work, in the case of low temperatures, and work at elevated heights in windy areas
- other such as, for example, landmines and unexploded ordinance such as bombs. The findings of archival and physical surveys should be included in the report and, if possible, the approximate positions recorded on the site plan. Furthermore, access roads may be an addressed in a similar manner
- project description such as, for example, the nature, scope and complexity of the project should be addressed in terms of the resulting general design and construction hazards and risks, and how they were addressed
- phases and programme:
  - responses to specific hazards and risks in terms of basements, including design and construction method statements, in the case of deep basements
  - temporary screening, overhead protection, signage and lighting required relative to the refurbishment of a shopping mall, including detailed specifications,

requirements and the annotation of site plans, floor plans and elevations, which should not be left to the discretion of the principal contractor (PC) and contractors

☐ furthermore, the cost engineer or quantity surveyor (QS) can then make informed inclusions in the bill of quantities (BoQ).

3.4.2 Client's considerations and management requirements

Client's considerations and management requirements relevant to the designer report include the following items which designers should confirm having taken cognisance of and/or responded to:

- provision of designers' designing for construction H&S documents, competencies, resources, processes and practices should be provided as indicated in the designer H&S specification
- community participation: designers should indicate how they have accommodated community participation in terms of the design, details and specification, and the implications thereof in terms of general skills, H&S awareness, knowledge and skills
- project H&S goals, such as reducing incidents and accidents, and improving H&S practices on projects, will lead to a fatality-, injury- and disease-free project, that is, zero harm in terms of the key H&S project objective. Critical designer objectives would then include practice commitment to construction H&S, in which case, the specific objectives should be, among others, the provision of a practice H&S policy and the provision of the practice's project designer philosophy statement
- H&S monitoring and review: designers need to indicate the process that was implemented during design and the process that will be implemented during construction, including the approach to address design-originated hazards and risks encountered during construction
- permit and authorisation requirements such as, for example, designer confirmation that on-site welding and other hot work, which could constitute a hazard and risk with respect to a client's on site activities, have been averted through design interventions such as details, specifications and methods of fixing
- fire prevention and protection such as, for example, designer confirmation that flammable materials have not been specified or limited, in the case of both brownfield and greenfield projects
- emergency procedures: designer confirmation of consideration of these in terms of planning and design in the case of brownfield projects, eg emergency exits and routes will not be compromised by the proposed alterations and/or additions and/or refurbishment

- site rules and other restrictions on contractors, suppliers and others, for example:
  - access arrangements to those parts of the site that continue to be used by the client, shift work, night work, restricted hours, dust and noise
  - confirmation of consideration of the maximum permissible size of components upon delivery, handling or vertical transportation
  - inclusion of prefabricated components and mitigation of onsite work
  - avoidance of dust and noise producing processes, and details and specification for temporary hoarding and screening
- activities on or adjacent to the site during the works, for example, construction work relative to the repair or replacement of live sewers, and the provision of design and construction method statements
- arrangements for liaison between parties in terms of, among others, routing and approval of specialist contractors' detailed shop drawings and temporary works in relation to the permanent works. The designers should confirm such arrangements in the form of a process flow diagram.

3.4.3 Environmental restrictions and existing on-site risks

Environmental restrictions and existing on-site risks should indicate the response to the delinked health hazards and safety hazards in the designer H&S specification.

Health hazards include:

- Asbestos: The results of the requisite archival and site surveys should be included, and the response in the form of encapsulation or removal should be indicated. Consideration must be given to the inclusion of the requisite items in the BoQ.
- Existing storage of hazardous materials such as, for example, fuel tanks in a tank farm. The designer(s) should confirm consideration thereof in the form of construction activities that have been avoided and the non-inclusion of flammable materials in the design, which may contribute to conflagration in the event of a fire.
- Contaminated land, in the case of brownfield and greenfield projects. The results of the requisite archival and site surveys should be included and the response should be indicated such as, for example, displacement versus augured piles.
- Health risks arising from client's activities, eg sewage treatment works and nuclear power plants. The designer(s) should indicate how they have mitigated the likely hazards and risks to construction workers in terms of design and construction method statements, including any engineering controls.

Safety hazards include:

- Adjacent land use that may include existing buildings or structures. Designers should confirm that the foundations thereof and related services will not be compromised by the proposed project's earthworks, and provide details of shoring in the form of raking shores. The designers should also indicate how any possible surcharge courtesy of the neighbouring building or structure during the project's earthworks has been accommodated in the design.

- Location of existing services and utilities, such as electricity, fire main, gas, sewage, stormwater and water, which may be underground, above ground or overhead, the latter being visually obvious. Designers should provide the findings of archival surveys, which would include an existing H&S file, existing site and services' plans and physical surveys. The site plan and respective services' plans should be annotated to indicate the positions/routes of such services and design and construction method statements indicating safe excavation procedures and practices should be provided.

- The ground conditions in terms of soil type(s) and related ground aspects should be annotated on the site plan, and design and construction method statements, including engineering controls, should be provided.

- Existing structures in terms of the degree of stability, in the case of elements that need to be retained and included in the proposed structure, such as a historical façade, or parts of a structure that will be demolished, should be addressed in the form of design and construction method statements. The processes to be followed in the case of temporary works for, say, any necessary dead shoring and possible accompanying raking shoring, should be prescribed. The treatment of fragile elements, components or materials, such as existing suspended timber floors, including the supporting structure and roof lights, should be clearly annotated on floor plans, roof plans and sections.

The health hazards and safety hazards, in terms of the environmental restrictions and existing on-site risks, should be addressed in a project register and categorised in terms of activities as per Table 8.1. A brief description of what the activity entails should be provided, along with the response to the hazard, the respondent in terms of the latter response, the future action relative to each respondent, the status of the future action and the related reference.

3.4.4 Significant design and construction hazards

The identification of the raw and residual hazards and related-risk ratings, and designer responses were addressed in Chapter 7. Table 7.3 provided a range of examples relative to various activities.

In addition to including the significant design and construction hazards in the designer report project hazard and risk register, as per Table 8.1, H&S hazards should be referenced on site plans, floor plans, elevations, details, structural drawings, reinforcement layouts and on the various schedules. Furthermore, where necessary, designers should prepare and include design and construction method statements in the designer report.

Table 8.1: Designer report project hazard and risk register

Activity	Description	Hazard	Response	Respondent	Future action	Status	Reference
Site clearance	Clearing bush	▪ Venomous snakes ▪ Thorn trees	Included in designer report	▪ Client ▪ PC	▪ Include in H&S specification ▪ Include in H&S plan	▪ Client to action ▪ Response awaited from PC	▪ Site plan ▪ Site plan
Earthworks	Excavations	Collapsible soil	Included in designer report	▪ Client ▪ QS ▪ PC	▪ Include in H&S specification ▪ Address in BoQ ▪ Include in H&S plan	▪ Client to action ▪ QS to action ▪ Response awaited from PC	▪ Base layout ▪ Trial hole results
Suspended RC slab construction	Erection of falsework	Handling, erecting & striking of falsework	Flat slabs (enable use of table forms)	None	None	None	Not required
	Post-tensioning	Injuries during stressing	Included in designer report	▪ Client ▪ PC	▪ Include in H&S specification ▪ Include in H&S plan	▪ Client to action ▪ Response awaited from PC	▪ Structural plan ▪ Post-tension cable layout
Masonry walling	Laying blocks	Manual handling of heavy blocks	Substituted with maxi bricks	None	None	Completed	Not required
Floor coverings	Laying vinyl floor sheeting	Exposure to irritants and toxins	Specified water-based adhesives as opposed to solvent-based adhesives	None	None	Completed	Not required
Site works	Kerbing	Handling heavy kerb stones	Specified extruded in situ kerbs	None	None	Completed	Not required
	Laying pavers	Cutting pavers due to herringbone pattern	Revised pattern to basketweave/parquet	None	None	Completed	Not required
	Asphalt road surfacing (hot mix)	Exposure to vapours and aerosols	Included in designer report (warm mix not accepted)	▪ Client ▪ PC	▪ Include in H&S specification ▪ Include in H&S plan	▪ Client to action ▪ Response awaited from PC	▪ Roadworks specification ▪ Sections
Landscaping	Planting cotton palms	▪ Weight ▪ Striking or crushing of workers	Included in designer report	▪ Client ▪ PC	▪ Include in H&S specification ▪ Include in H&S plan	▪ Client to action ▪ Response awaited from PC	▪ Landscaping layout ▪ Plant schedule

3.4.5 The H&S file

The **H&S** file should be addressed in the designer report, and the following constituting issues should be addressed:

- location and nature of utilities and services, including the design interventions relative to the mitigation of accidental discovery of new services, such as high voltage cables that have been protected by concrete slabs and identified by black and yellow chevron tape within trenches beneath backfill
- design criteria such as, for example, design loadings, including calculations, of suspended reinforced concrete slabs within the context of mitigating overloading during the occupancy of the building or structure
- construction methods and materials used such as, for example, post-tensioned reinforced concrete slabs and self-cleaning glazing
- potential hazards included in the structure such as, for example, post-tensioned cables and the indication of their positions on, say, the soffit of the related slabs
- record of hazardous processes such as, for example, removal and/or encapsulation of ACMs
- equipment and maintenance facilities such as, for example, window cleaning gondolas
- maintenance procedures and requirements such as, for example, escalators, lift installation and window cleaning gondolas
- manuals (operating and maintenance) for plant and equipment such as, for example, window cleaning gondolas.

4. Review Questions

- What is the purpose of the designer report?
- What are the pre-requisites to realise an informed designer report?
- Based upon your experience, in general, are designer reports adequate?

5. Review Exercises

- Compile a designer report for a project based upon your design HIRA.
- Compile a range of design and construction method statements.

CHAPTER 9

Contractor Health and Safety Specification

John J Smallwood

1. Introduction

In terms of Regulation 5, 'Duties of client', 5(1)(b) of the Construction Regulations, 2014, clients are required to prepare an H&S specification based on the baseline risk assessment (BRA) and, thereafter, 5(1)(f) requires the client to include the H&S specification in the tender documents.

The Construction Regulations define an **H&S specification** to mean 'a site, activity or project specific document prepared by the client pertaining to all health and safety requirements related to construction work'. Given that designers are required to provide clients with a report following the receipt of the designer health and safety (H&S) specification and the required contents of the report, the H&S specification provided to the designers should be referenced as 'designer', and that the H&S specification must be included in the tender documents and should be referenced as 'contractor'. Although the Construction Regulations do not differentiate between the two obvious versions of the H&S specification, it is notable that the Scope of Work for Categories of Registration of the Project and Construction Management Professions, 2019 (Scope of Work) for construction health and safety agents (CHSAs) records: '3.7 Identify and implement precautions necessary for construction project health and safety control and update the construction project tender health and safety specifications' relative to Stage 3, design development, and '4.3 Finalise construction project tender health and safety specifications and integrate with procurement documentation' relative to Stage 4, tender documentation and procurement', according to The Scope of Work for Categories of Registration of the Project and Construction Management Professions, 2019 (Scope of Work).

2. Intention/Rationale

As stated in Chapter 2, H&S specifications are not intended to be a regurgitation of the Construction Regulations but to serve as a guide in terms of the hazards and risks, and the client's H&S requirements the principal contractor (PC) and contractors need to address relative to a project, that is, project specific. Clients must base the H&S specification on the BRA, which should inform with respect to the indirect and direct project-wide hazards and risks, and designers should include the raw or residual hazards and risks remaining after design hazard identification and risk assessments (HIRAs) in the designer reports, which, in turn, should be included in the contractor H&S specification.

3. Evolution

The Scope of Work for CHSAs records: '4.3 Finalise construction project tender health and safety specifications and integrate with procurement documentation.'

It is imperative that the evolution of the contractor H&S specification is informed by the preceding processes and documents: BRA → designer' H&S specification → design HIRA → designer report → contractor H&S specification → H&S plan → H&S file because failure to do so will militate against the intention of the sequence (process), namely sequential reduction of hazards and risks.

4. Contents

The client provided contractor H&S specification should address the following:
- project details
- client's considerations and management requirements
- environmental restrictions and existing on-site hazards and risks
- significant design and construction hazards
- H&S file.

Environmental issues, which, although they should be addressed in the environmental impact assessment, should also be addressed in the contractor H&S specification as they may well be related to H&S such as, for example, fires and the removal and disposal of asbestos-containing material (ACM).

4.1 Project details

Project details include the following information, namely
- project location including:
  - access to the project location in terms of infrastructure, such as airports, harbours, railway routes above and below ground, rivers and roads, including conditions, load restrictions and restrictions with respect to transportation of specific goods that may impact on the supply of materials and plant and equipment to the project such as, for example, wind turbine masts and blades

- adjacent public thoroughfares, such as roads, sidewalks and above ground railway routes in terms of considerations and/or restrictions such as, for example, protection of the public while using such sidewalks and façade retention relative to a redevelopment project involving demolition
- access to the site in the form of entrances and/or exits from public thoroughfares in terms of restrictions such as, for example, tight turning circles, traffic volumes and peak traffic periods that may affect, among others, the length of delivery vehicles that can enter and exit the site
- consideration of availability of space for site accommodation, plant and equipment, lay down areas, the delivery of materials to the site and the blocking of public thoroughfares while offloading. The client may be able to provide areas for site accommodation, including welfare facilities in the case of brownfield projects
- fauna, flora and related issues such as, for example, respect for dangerous animals in the form of carnivores/predators, crocodiles, malaria, scorpions, snakes and tortoises. Furthermore, protected species in terms of flora, such as fynbos in the case of the Table Mountain National Park, which is susceptible to concrete and other cementitious materials and fires. Given the latter, naked fires and smoking are prohibited
- viruses and pandemics such as, for example, Covid-19, HIV and AIDS, malaria and sleeping sickness;

Figure 9.1: Mist in the vicinity of the upper station, Table Mountain Aerial Cableway station project, (Deacon, 1997)

- □ existing services and utilities such as, for example, electricity, fire main, gas, sewage, stormwater and water, including access thereto
- □ socioeconomic issues, such as the 'construction mafia', SMMEs, crime and vandalism
- □ weather such as, for example, lightning, mist, precipitation (including hail), temperature (including humidity and wind chill factor) and wind (including prevailing direction)
- □ other such as, for example, the likelihood of landmines and unexploded ordinance such as bombs
- project description in terms of the nature, scope and complexity such as, for example, new build in the form of a hotel, refurbishment of a hotel or refurbishment of a shopping mall, including alterations and additions. Characteristics include:
  - □ erf size
  - □ current floor area
  - □ additional floor area
  - □ storeys
  - □ capacity, such as 300 bedrooms in the case of a hotel
- phase and programme requirements such as, for example, separate phases for the basement and the structure and continued use of a shopping mall during refurbishment, including tenants changing stores
- extent and location of existing records and plans, which will provide information about the position of services, existence of post-tensioned slabs and the existence of ACMs.

4.2 Client's considerations and management requirements

Client's considerations and management requirements include:
- provision of the PC's:
  - □ personnel's qualifications, statutory registration and H&S competencies
  - □ H&S policy, management system and programme
  - □ general and H&S project organogram
  - □ general resources
- industry 4.0 technologies such as drones and sensors, where the PC should give advice in the H&S plan regarding the use thereof relative to H&S
- client contractor H&S management programme, where repeat clients may have such a programme in place and require the PC and contractors to participate therein, especially in the case of a brownfield project
- H&S education and training such as, for example, PC's and contractors' personnel are required to attend H&S induction presented by the client, especially in the case of a brownfield project
- workers' compensation insurance letters of good standing

- participation in a multi-stakeholder partnering process, which includes H&S
- community expectations in terms of participation in the project, which has implications in terms of general construction knowledge and skills, and the H&S awareness, knowledge and skills required to, among others, work in excavations or at heights
- project H&S goals are likely to include reducing incidents and accidents, and the improvement of H&S practices on projects, therefore, a key project objective in terms of H&S is likely to be a fatality-, injury- and disease-free project, such as zero harm, where contractors need to indicate how they will contribute thereto
- H&S education and training such as, for example, PC's and contractors' personnel are required to attend client H&S induction
- H&S monitoring and review such as, for example, the PC's leading and trailing H&S indicators, HIRA processes, available SWPs and H&S inspection regime, including the personnel responsible for the inspections
- permit and authorisation requirements as a result of the client's onsite activities on a brownfield project such as, for example, avoidance of on-site welding and other hot work
- fire prevention and protection because if client's and/or users' activities during occupancy
- emergency procedures, in the case of brownfields projects such as, for example, alterations and additions or refurbishment
- site rules and other restrictions on contractors, suppliers and others such as, for example, access arrangements to those parts of the site that continue to be used by the client, workdays per week, shift work, night work, restricted hours, limiting of in situ and on-site work, dust and noise
- requirements relative to the transportation of workers to and from site as a contribution by the client to contribute to reducing MVA-related fatalities and injuries
- activities on or adjacent to the site during the works such as, for example, construction work relative to the repair or replacement of live sewers
- arrangements for liaison between parties. This requirement could relate to the routing and approval of specialist contractors' detailed shop drawings, or temporary works in relation to the permanent works, such as the attachment of tower crane stays to the permanent structure
- submission of a project H&S closeout report or contribution thereto.

4.3 Environmental restrictions and existing on-site hazards and risks

Environmental restrictions and existing on-site hazards and risks should address the health hazards and safety hazards.

4.3.1 Health hazards

Health hazards include:

- asbestos, in the case of brownfield projects, especially those entailing alterations and additions or refurbishment
- existing storage of hazardous materials such as, for example, fuel tanks in a tank farm
- contaminated land, in the case of both brownfield and greenfield projects, which is likely to affect activities
- health risks arising from the client's activities such as, for example, sewage treatment work and nuclear power plants.

4.3.2 Safety hazards

Safety hazards include:

- Adjacent land uses, which include existing buildings or structures, the foundations and related services of which will constitute an issue should the proposed project involve the construction of a basement or, for that matter, standard excavations relative to column, and retaining wall bases and footings. Furthermore, the consequences of possible structural failure need to be considered.
- Location of existing services and utilities such as electricity, fire main, gas, internet, sewage, stormwater and water, which may be underground, above ground or overhead
- Ground conditions, such as the soil type, which has stability implications, ground water level, possible ingress of water, old mine workings, related possible contamination in the case of harbouring of leaking fuel tanks, or other hazardous materials, underground obstructions such as foundations and possible subsidence.
- Existing structures such as, for example, degree of stability, in the case of elements that need to be retained and included in the proposed structure such as a historical façade, or parts of a structure that will be demolished. Fragile elements, components or materials, such as existing suspended timber floors, including the supporting structure and roof lights need to be considered.

4.4 Significant design and construction hazards

This section is intended to advise the PC and contractors about the residual hazards, including raw hazards that cannot not be mitigated, ie in addition to the environmental restrictions and existing on-site risks in the form of health hazards and safety hazards as addressed in section 4.3. Table 9.1 comes from the designer report project hazard and risk register, and presents the activities, descriptions thereof and the hazards, along with the responses thereto and the respondents, the future actions required, the status and where the hazards are referenced.

Table 9.1: Contractor H&S specification project hazard and risk register

Activity	Description	Hazard	Response	Respondent	Future action	Status	Reference
Site clearance	Clearing bush	▪ Venomous snakes ▪ Thorn trees	Included in designer report	▪ Client ▪ PC	▪ Included in H&S specification ▪ Include in H&S plan	▪ Completed ▪ Response awaited from PC	▪ Site plan ▪ Site plan
Earthworks	Excavations	Collapsible soil	Included in designer report	▪ Client ▪ QS ▪ PC	▪ Included in H&S specification ▪ Included in BoQ ▪ Include in H&S plan	▪ Completed ▪ Completed ▪ Response awaited from PC	▪ Site plan ▪ Base layout ▪ Trial hole results
Suspended RC slab construction	Erection of falsework	Handling, erecting & striking of falsework	Flat slabs (enable use of table forms)	None	None	None	Not required
	Post-tensioning	Injuries during stressing	Included in designer report	▪ Client ▪ PC	▪ Included in H&S specification ▪ Include in H&S plan	▪ Completed ▪ Response awaited from PC	▪ Structural plan ▪ Post-tension cable layout
Masonry walling	Laying blocks	Manual handling of heavy blocks	Substituted with maxi bricks	None	None	Completed	Not required
Floor coverings	Laying vinyl floor sheeting	Exposure to irritants and toxins	Specified water-based adhesives as opposed to solvent-based adhesives	None	None	Completed	Not required
Site works	Kerbing	Handling heavy kerb stones	Specified extruded in situ kerbs	None	None	Completed	Not required
	Laying pavers	Cutting pavers due to herringbone pattern	Revised pattern to basketweave/parquet	None	None	Completed	Not required
	Asphalt road surfacing (hot mix)	Exposure to vapours and aerosols	Included in designer report (warm mix not accepted)	▪ Client ▪ PC	▪ Included in H&S specification ▪ Include in H&S plan	▪ Completed ▪ Response awaited from PC	▪ Roadworks specification ▪ Sections
Landscaping	Planting cotton palms	▪ Weight ▪ Striking or crushing of workers	Included in designer report	▪ Client ▪ PC	▪ Included in H&S specification ▪ Include in H&S plan	▪ Completed ▪ Response awaited from PC	▪ Landscaping layout ▪ Plant schedule

4.5 The H&S file

Given that the designers, PC and contractors need to contribute thereto, the H&S file contents in the respective H&S specifications are similar, among others:

- as-built drawings and plans
- design criteria such as, for example, design loadings and calculations
- construction methods and materials used such as, for example, post-tensioned reinforced concrete slabs
- potential hazards included in the structure such as, for example, post-tensioned cables
- record of hazardous processes such as, for example, removal or encapsulation of ACMs, in which case the workers' names and identification numbers should be documented
- equipment and maintenance facilities
- maintenance procedures and requirements
- manuals (operating and maintenance) for plant and equipment
- location and nature of utilities and services, including an as-laid site plan.

5. Review Questions

- What is the purpose of the contractor H&S specification?
- What are the pre-requisites to realise an informed contractor H&S specification?
- Based upon your experience in general, are contractor H&S specifications informative?

6. Review Exercises

- Compile a contractor H&S specification for a project based upon your BRA and designer report therefor.
- Conduct a review of a contractor H&S specification.

CHAPTER 10

Pricing for Construction Health and Safety

John J Smallwood

1. Introduction

In terms of Construction Regulation 5(1)(g), clients are required to ensure that potential principal contractors (PCs) have made provision for the cost of health and safety (H&S) in their tenders. Furthermore, contractors must resource H&S to comply with H&S legislation and regulations, and their own H&S standards.

However, pricing H&S constitutes a challenge for several reasons:

- most projects are unique, despite many common activities
- estimates must be accurate
- cost of H&S data must be available
- contractors committed to H&S may be competing with contractors that are less committed to H&S
- contractors have varying levels of expertise, H&S included
- emerging contractors and SMMEs should be guided.

Given that the Construction Regulations do not require a pre-tender H&S plan, it is not possible for the client to relate the PC's provision for the cost of H&S to the H&S plan, which, in theory, is only reviewed before the commencement of construction.

2. Cost of H&S Data

Historically, researchers have encountered difficulty in terms of determining the cost of H&S for several reasons. Projects differ in terms of:

- type of building or structure
- type of structural frame
- number of storeys
- area (m^2) per storey
- extent of compartmentation per storey
- enclosing fabric
- plant and equipment and temporary works required
- extent of contributors in the form of subcontractors or trade contractors.

A cost of H&S study conducted on a small sample of best practice H&S general contractors (GCs) realised notable findings namely 77.8 per cent of the GCs responded that they allocated H&S costs on projects to an identifiable cost centre/code, while 22.2 per cent said that they did not. Respondents were then requested to indicate how they allocated such costs in terms of 16 H&S programme elements. The costs of eight of the 16 elements (that is, 50 per cent of the elements) were allocated to an H&S cost centre/code by more than 50 per cent of the respondents, whereas the costs of six of the 16 elements (namely 37.5 per cent of the elements) were allocated elsewhere by more than 50 per cent of the respondents. In terms of the percentage that the PC cost of H&S contributes to the total project cost, two GCs recorded a percentage of 0.5 per cent and 3.0 per cent, while six GCs indicated percentage ranges, with three GCs recording a range of $> 0 \leq 1$ per cent and three recording a range of $> 1 \leq 2$ per cent (Smallwood, 2004).

A study conducted on GC members of the Kwazulu-Natal Master Builders Association who achieved places in a regional H&S competition, determined that the mean cost of H&S as a percentage of the tender cost was estimated to between 1 per cent and 1.6 per cent of the project cost.

3. Mechanisms for Pricing Construction H&S

What do the findings of South African research indicate? A study set out to determine the concurrence between a range of statements. The mean scores, reported in parentheses, were between 1.00 and 5.00. Mean scores closer to 5.00 indicate a high level of concurrence. The findings were as follows:

- 'Contract document enabled financial provision for H&S promotes H&S' (4.36)
- 'A detailed H&S section should be included in the preliminaries' (4.27)
- 'A provisional sum should be provided for H&S in the preliminaries' (3.64).

Given that figures are dependent on a range of variables, the inclusion of a unit item in the preliminaries section of a bill of quantities for H&S would not be ideal for at least two reasons. First, it would not constitute scientific estimating on the part of contractors; and secondly, clients would not be able to ensure that potential PCs have made provision for the cost of H&S in their tenders.

Similarly, the inclusion of a provisional sum would not be ideal because, first, the sum would invariably be based on a percentage of the project cost estimate; secondly, the administration of payment on a proof of expenditure basis would be time consuming; and thirdly, H&S-related purchases would then only be debited to a single project, despite residual value.

A schedule of H&S preliminaries items based on an informed H&S specification and a designer report is best because it will enable a scientific approach to pricing H&S and an informed review by the client and other members of the tender adjudication team of the provision for the cost of H&S. This contention is supported by the recording in the Scope of Work for Categories of Registration of the Project and Construction Management Professions, 2019 (Scope of Work) of '3.11 Assist the cost consultant with detailed information for initial construction project health and safety cost estimates/budgets' relative to Stage 3, design development, and '4.8 Assist the cost consultant in the finalisation of the construction project health and safety cost estimate/budget' relative to Stage 4, tender documentation and procurement, in the CHSA's Scope of Work. However, the latter should be expedited in the initial stages of Stage 4.

After the tender adjudication and awarding of the project, reconciliation of the PC's H&S plan relative to the PC's provision for the cost of H&S will provide further insights into the adequacy of the provision made. The award of projects prior to the discussion and approval of the PC's H&S plan amplifies the need for the inclusion of a pre-tender H&S plan with the tender submission as a requirement in the Construction Regulations.

4. Schedule of Preliminaries Items

A study conducted on medium- to large-sized GC members of the East Cape Master Builders Association and members of the Eastern Cape Province Chapter of the Association of South African Quantity Surveyors investigated a range of issues relative to financial provision for construction H&S. Respondents were required to indicate the importance of the inclusion of 39 items in an H&S preliminaries section. The mean score of all the responses received from the general contractors and quantity surveyors was > 4.00, the midpoint of the mean scores range between 1.00 and 7.00, which indicates that the inclusion preliminaries is more than important, as opposed to less than important.

Table 10.1 presents a schedule of preliminaries items. The key issue with respect to pricing for construction H&S is that many items have substantial financial implications. A further issue is that consulting engineers and QSs are likely to want to limit their liability and, furthermore, they may be unable to estimate or quantify the number of items, as in the case of medicals examinations. Although they would not be able to determine the number of medical examinations the PC has priced for at tender stage, the risk is transferred to the PC, who must estimate and make a global allowance for medical examination. However, an estimate of the size of the workforce will provide guidance on how many are likely to be conducted.

There are several items that are not included in the schedule because they are temporary works-related, namely:
- fencing or hoarding
- protected public and site walkways

- catch platforms
- scaffolding, suspended scaffolding and special scaffolding.

The compilation of an H&S plan, fall protection plan, H&S file and method statements requires expertise, which may need to be outsourced. If the expertise exists in-house, then contractors can, in theory, recover the 'expense' of employing such an expert through their percentage provision for overheads in the case of a successful tender.

Risk assessments should be conducted by the people who are going to undertake the activity and, to a degree, safe working procedures; however, external expertise may be required to facilitate the processes. Production workers should be involved in hazard identification and risk assessment (HIRA), which means that they are not producing while doing so and, therefore, the cost of this involvement should be allowed for.

Design of temporary works, permanent works, engineering design and/or certification and environmental measurement are specialist activities, and are likely to involve outsourcing and, therefore, they are non-inclusion as an overhead expense. These activities can be readily identified and quantified.

Environmental measurement is more challenging and may entail measurement of dust particles, lighting levels and noise; however, it is possible to develop an informed estimate. Medical examination are more challenging because employers should absolve themselves of any potential liability for occupational diseases incurred by employees during prior employment. Given that baseline medical examination are imperative for permanent and contract employees, the question arises whether the cost should accrue to the construction business overheads or the project on which they are deployed. Medical examinations are a requirement for plant operators.

In terms of the Regulations for Hazardous Chemical Agents, 2021, medical surveillance refers to the overall monitoring of employees to identify changes in their health status as a result of exposure to certain hazardous chemical agents (HCAs). Monitoring activities are not limited to testing but include the monitoring and analysis of the individual and group outcome data, including historical data, derived from the medical testing. Medical testing as an aspect of medical surveillance, therefore, involves the use of interviews, questionnaires, and standard clinical assessments to detect the presence of adverse health effects. This can also include tests such as spirometry (lung function), radiography in the form of chest x-rays and laboratory tests in the form of full blood counts. Biological monitoring provides an additional means of assessing the exposure to an HCA by measuring metabolites of the HCA or other similar markers of exposure. It is likely to be challenging to estimate the costs related to medical surveillance and biological monitoring because of the need to include PC and subcontractor employees in the budget, however, an erudite PC that has consulted an occupational health or occupational medical practitioner should be able to make an informed estimate.

While H&S education and training can readily be identified, however, the extent thereof will require deliberation because it may include more than HIRA and safe work procedures training. Other training could include supervisor H&S, H&S representative, first-aider and firefighting training. However, it is possible to obtain quotations to deliver various types of education and training.

Provision for one or more H&S officers is relatively straight forward. H&S representatives, on the other hand, are elected by the workforce and, therefore, their H&S-related contributions extend beyond their production function, so an estimate should be made to determine the related cost arising from H&S representative inspections, attendance of H&S committee meetings and participation in H&S investigations. It is possible to use the minimum legislative requirements as guidelines.

H&S committees, H&S meetings and H&S inspections involve a range of PC employees, most of who are costed relative to overheads, however, some may be involved in production and, therefore, lost production should be budgeted for. H&S administration is overarching and entails contributions from management, supervision, H&S practitioners, administrators and production workers. Once again, the related cost should be identified and budgeted for.

First aid includes the provision of the first aid room and ancillary facilities, although this may be included in the site accommodation item in the standard preliminaries. However, the furnishings and fittings, such as a bed, stretcher(s), first aid kit and first-aider, should be priced into the H&S preliminaries.

The provision for personal protective equipment (PPE) should be based on the standard PPE per trade per worker, the challenge will be the likely number of workers deployed throughout a project. However, specialist PPE, such as personal fall arrest protection, respirators and other breathing apparatus and special suits, in the case of asbestos removal, that workers will have to be priced and included in the estimate.

Pricing of the transport of workers is dependent on a range of issues, among others:

- legislation
- organisation policy
- possible client requirements
- location of the project.

The challenge will be determining the likely number of workers deployed throughout a project.

H&S induction may be delivered by the client, depending on the client's requirements, in which case, provision should be made for lost production. Contractor delivered induction will require a similar provision, in addition to provision for possible presentations by a consultant, if not presented by the contractor's staff, such as the construction health and safety officer or the construction health and safety manager. The challenge in either case will be determining the likely number of workers deployed throughout a project.

Site access control may be prescribed by client requirements. However, it may, to a degree, be included in the site accommodation item of the standard preliminaries. Site access control may not only include the physical entrance but also a security guard and a range of security access control equipment and information and communications technology, which may include access cards.

The following are the seven categories of H&S-related signage:
1. mandatory
2. prohibition
3. warning
4. danger
5. fire safety
6. emergency information
7. restriction.

The categorisation of signage, along with an informative site layout and other documentation such as a fire safety plan, will facilitate pricing for signage.

Emergency planning entails a range of considerations, including possible client requirements. The development of the emergency plan, itself, may need to be outsourced or, if the expertise exists, developed in-house. Another consideration would be a simulated emergency exercise, which may have cost implications in terms of lost production on the project.

Facilities in the form of toilets, wash hand basins, showers, mess rooms and change rooms, require deliberation in terms of the likely number of workers on site and the choice between chemical and flush toilets. Contractors may have mobile ablution facilities, which could be connected to the municipal system. Pricing must, therefore, include:
- the installation, whether mobile, in situ or temporary
- maintenance
- consumables, such as toilet paper and paper towel should hot air dryers not be provided
- an estimate for municipal charges.

The requirement for living accommodation depends on the location of a project, the key issues being the number of workers that need to be accommodated, and the nature and scope of the accommodation. Overall, Regulation 30, 'Construction employees facilities', of the Construction Regulations prescribes the minimum requirements.

Figure 10.1: Workers change room, shower and lockers (left), and workers' mess area (right), Max 4 project, Lund, Sweden (Smallwood, August 2012)

Pricing for Covid-19 compliance and/or other medical protocols post-Covid includes provision for:

- site entrance testing station(s) including personnel testing
- infrared thermometers
- hand sanitisers
- masks
- disposal bins and waste disposal costs
- sanitising points
- sanitising of areas, tools and vehicles
- related training.

Other requirements can be accommodated by the provision for facilities.

Flammable goods require dedicated storage, in accordance with Construction Regulations Regulation 25, 'Use and temporary storage of flammable liquids on construction sites'. Furthermore, the likely range and volume of such goods and related storage requirements, such as racking, should be considered. PCs may have mobile flammable goods stores, which could avoid repeat establishment costs.

Fire precautions and prevention are prescribed by Regulation 29, 'Fire precautions on construction sites', of the Construction Regulations. Ideally, however, PCs should develop a fire safety method statement for projects at the pre-tender stage, in conjunction with a site layout and a fire safety plan, for inclusion in the pre-contract H&S plan. It is likely that fire extinguishers, which need to be serviced at six-monthly intervals, will predominate. Pricing should allow for firefighting exercises to ensure a degree of competence should a fire break out on a project.

Regulation 24, 'Electrical installations and machinery on construction sites', of the Construction Regulations are prescriptive in terms of temporary electrical installations. This regulation is complemented by the wiring code set out in South African National Standards (SANS) 10142-1. In essence, item 7.4 of this code, 'Construction and demolition site installations', requires that an electrical contractor should be engaged to undertake a temporary electrical installation and the maintenance thereof.

Guarding and barricading, which constitutes a major cost, can be approached on an opening and lineal meter basis and fairly accurately measured.

Liaison with client and/or H&S agent by the PC during the project execution may include specific client requirements and, even though the liaison may be courtesy of PC staff that are included in the project overheads, ancillary costs arising therefrom may need to be priced for.

Identification of employees may be a client requirement, in which case, the challenge is to estimate the likely number of workers and other employees on a project. Various forms of personal identification are available that could be linked to an access control system.

The location of services and utilities that constitute a hazard in terms of the baseline risk assessment and the contractor H&S specification requires their likely position to be indicated by the client, their exact location should be confirmed before excavation activities commence and, therefore, the process should be included in the H&S preliminaries. Furthermore, the process may need to be outsourced because of its specialist nature.

Containment areas for hazardous work may be required in the case of the removal of asbestos-containing materials (ACMs) to prevent the spread of the related fibres. Such areas may require sealed hoarding, mechanical extraction and a sterilisation station to ensure that workers do not spread fibres that may have become attached to their protective suits. In the case of ACMs, a registered asbestos contractor must undertake the removal, transport and disposal of such materials and therefore be able to provide a quotation to be included in the final tender or bid.

Rubble chutes could be included in the standard preliminaries or in the H&S preliminaries. The cost of these it could be argued should be included in the standard construction activity rates.

Removal of hazardous material may be measured in a trade such as demolition, or in the H&S preliminaries. Given that it is a specialist activity, it will be outsourced. In the case of ACMs, in addition to the appointment of a registered asbestos contractor, the following are key requirements of the HCA Regulations, depending on the type of asbestos:

- an asbestos risk assessment is required for either asbestos repair or removal work
- appointment of an approved inspection authority
- a plan of work
- air monitoring
- medical surveillance
- PPE
- labelling and signage
- prescribed transportation and disposal.

Table 10.1: Schedule of possible H&S preliminaries items

Item description	Quantity	Rate	Amount
H&S plan	Item		
Fall protection plan	Item		
H&S file	Item		
Risk assessment	No.		
Safe working procedures	No.		
Method statements	No.		
Design of temporary works	Item		
Design of permanent structures	Item		
Engineering design and/or certification	Item		
Environmental measurement	Item		
Medicals	No.		
Medical surveillance	No.		
Biological monitoring	No.		
H&S education and training	Item		
H&S Officer(s)	No.		
H&S Representative(s)	No.		
H&S Committee	Item		
H&S meetings	No.		
H&S administration	Item		
H&S inspections	No.		
First aid	Item		
PPE	Item		
Transport of employees	Item		
H&S induction	Item & No.		
Site access control	No.		
H&S signage	No.		
Emergency planning	Item		
Facilities:			
Toilets	No.		
Wash hand basins	No.		
Showers	No.		
Mess rooms	Item		
Change rooms	Item		
Living accommodation	Item		
Covid-19 compliance	Item		
Storage for flammable goods	Item		
Fire precautions and prevention	Item		
Temporary electrical installations	Item		

$\longrightarrow$

Item description	Quantity	Rate	Amount
Guarding and barricading	Item		
Liaison with client and/or H&S agent	Item		
Identification of employees	No.		
Location of unknown services	Item		
Containment areas for hazardous work	Item		
Rubble chute(s)	No.		
Removal of hazardous material	Item		

5. Review Questions

- Was the requirement that clients must ensure that potential PCs have made provision for the cost of H&S in their tenders as per the Construction Regulations, justified? A philosophical and pragmatic response is required.
- To what extent does competitive tendering promote construction H&S? A philosophical and pragmatic response is required.
- Would a pre-tender H&S plan promote construction H&S? A philosophical and pragmatic response is required.

6. Review Exercises

- Prepare a contractor estimate for H&S on a project by referring to the related-project documentation. This exercise may require a self-evolved contractor H&S specification and H&S plan.
- Compile a 750-word article titled *The link between the client baseline risk assessments, designer H&S specification, designer report and contractor H&S specification in terms of provision for the cost of H&S by the PC and contractors.*

Bibliography

Smallwood, JJ. 2004. Optimum cost: The role of health and safety (H&S). In: JJP Verster (ed) *Proceedings of the International Cost Engineering Council 4th World Congress*, Cape Town, South Africa (17–21 April 2004).

Smallwood, JJ. 2011. Financial provision for health and safety (H&S) in construction. In: *Proceedings of CIB W099 Conference 'Prevention: Means to the End of Safety and Health Injuries, Illness and Fatalities'*, Washington D.C., United States of America (24–26 August 2011).

Smallwood, JJ & Emuze, F. 2014. Financial provision for construction health and safety (H&S). In: DC Lacouture, J Irizarry & B Ashuri (eds) *Proceedings of the Construction Research Congress: Construction in a Global Network*, Atlanta, Georgia, United States of America (19–21 May 2014): 1881–1890.

Smallwood, JJ & Malan, R. 2018. Financial provision for construction health and safety. In: J Shiau, V Vimonsatit, S Yazdani & A Singh (eds) *Proceedings Fourth Australasia and South-East Asia Structural Engineering and Construction Conference 'Streamlining Information Transfer between Construction and Structural Engineering'*, Brisbane, Australia (3–5 December 2018).

CHAPTER 11

Construction Health and Safety Management Plan

Theo C Haupt

1. Introduction

Needless to say, effective planning is an essential prerequisite for effective construction health and safety (H&S) management. The adage of 'plan your work and work your plan' holds true. This important aspect was recognised and reinforced in the Construction Regulations, 2014 that were promulgated by the Minister of Labour on 7 February 2014 under section 43 of the Occupational Health and Safety Act 85 of 1993, which underwent a major revision about 11 years after they were first introduced on 18 July 2003. In terms of these regulations, all contractors must specifically provide and demonstrate to a construction client a suitable, sufficiently documented and coherent project- and site-specific construction H&S management plan based on the documented H&S specifications of the client. This management plan must be applied from the date of commencement and for the duration of the construction work, and must be reviewed and updated by the principal contractor (PC) as the work progresses on the construction project.

2. Requirements of a Construction H&S Plan

The regulations, therefore, have very explicit requirements that must be met when compiling a construction H&S plan. These requirements include, inter alia, the following:

- The plan must demonstrate the capability of the contractor to deliver a project without exposing the client, during the execution of the project, to any of the documented H&S challenges identified in the H&S specifications provided by the client at the tender stage.
- The plan must be related to a specific project and to a specific construction site.
- The contents of the plan must be discussed and approved after negotiations with a client before any construction work commences on the applicable site to ensure that it primarily addresses, completely and comprehensively as far as reasonably practicable, the challenges and issues identified in the H&S specifications.

- The plan must also include details of how the contractor will ensure that construction H&S is considered, generally, on all aspects of the project as the work proceeds.
- The plan must be a 'living' document that has to reviewed and updated regularly during the execution of the project.

It must, therefore, become blatantly obvious that a 'one-size-fits-all' approach or commercially available 'off-the-shelf standard construction H&S plans' would not meet the requirements of the regulations and are, in reality, inappropriate for use in the construction industry.

Appropriate H&S plans provide an integrated approach to the management of construction H&S on specific projects and sites. They set out the H&S goals of the project and explain how H&S issues will be managed during the execution of the project to control key project and site risks. While the requirements of the regulations might be considered onerous, sight must never be lost of their intention, which is to save lives, prevent injuries and disease and improve the quality of life of everyone involved in the construction project. The level of detail in the H&S plan must be proportionate to the risks involved in the particular project and on the particular site.

3. Considerations when Compiling a Construction H&S Plan

In general, the following need to be considered when compiling the construction H&S plan:
- identification of the key issues that must be addressed on the site through an effective hazard identification and risk assessment (HIRA) process
- development of action plans as a response to issues identified
- incorporation of relevant method statements, safe working or operating procedures, codes of practice and standards
- documentation of procedures for inspection, audit and review of the plan throughout the project duration.

4. Content of a Construction H&S Plan

It is necessary to approach the development of the H&S plan by taking into account the pre-tender and construction phase considerations as each phase involves different aspects of H&S for consideration. While there are no regulatory or prescribed requirements for a construction H&S plan, it should at least include the following:
- A construction H&S policy, which is a statement of the commitment by the management of the organisation to maintain the highest levels of H&S management and their involvement in the process.

- Project- and site-specific construction H&S goals, which describe the commitment of an organisation's management in more specific and measurable terms that are challenging and yet realistically achievable during the project and on site.
- The roles and responsibilities of project and site managers, supervisors, H&S staff and all workers.
- Discipline policy and procedures that provide details of consequences for failing to comply with the provisions of the H&S plan during the project and on site.

Example 11.1: Example of a discipline policy and procedures

CONSTRUCTION HEALTH AND SAFETY DISCIPLINE POLICY AND PROCEDURES

CHRISTOPHER BUILDING CONTRACTORS

Christopher Building Contractors believe that the health and safety way is the right way. Therefore, every construction worker is expected to know and adhere to our health and safety rules, as well as those prescribed by law. Failure to do so will result in the following disciplinary action, namely:

- Termination
  Christopher Building Construction reserves the right to immediately terminate any worker who deliberately and blatantly endangers themselves, other fellow workers or property as a result of unsafe attitudes and behaviour.

- Verbal warning (first violation)
  Non-blatant, impulsive and unpremeditated first violations will result in a verbal warning, accompanied by an appropriate record of the violation in the worker's personnel file.

- Written warning (second violation)
  Non-blatant, impulsive and unpremeditated second violations will result in a written warning, a copy of which will be placed in the worker's personnel file.

- Suspension (third violation)
  Non-blatant, impulsive and unpremeditated third violations will result in suspension without pay for a period of one week (5 working days).

- Dismissal (final violation)
  The next violation after a suspension will result in immediate termination of employment.

- Minor violations may be removed from the worker's personnel file at the rate of one violation removed for every six months of uninterrupted work without a violation and the completion within that same period of time of construction health and safety training for eight hours undertaken in the unpaid time of the worker.

Adapted from Goetsch, 2013

- Procedures for conducting regular inspections and audits.

Example 11.2: Example of an H&S inspection checklist

CONSTRUCTION HEALTH AND SAFETY INSPECTION CHECKLIST	
CHRISTOPHER BUILDING CONTRACTORS	

Project: ....

Date: ....

Mark with an 'X' only those areas where problems exist, attach an explanation of the hazardous condition that exists and what mitigation intervention should be introduced.

Problematic area	Comment
Housekeeping	
Personal protective equipment	
Noise	
Illumination/lighting	
Ventilation	
Fire protection/prevention	
Signage and barricades	
Material handling, storage, disposal	
Electrical issues	
Motor vehicles	
Plant and equipment	
Hand/power tools	
Concrete/brickwork	
Steel erection	
Demolition/blasting	
Stairs and ladders	
Scaffolding	
Vibration	

Signature:

- Details of the procedures for accident and incident investigations.

Example 11.3: Example of an H&S accident investigation statement

CONSTRUCTION HEALTH AND SAFETY INVESTIGATION STATEMENT

CHRISTOPHER BUILDING CONTRACTORS

Supervisors are required to conduct a detailed investigation whenever an accident occurs in their area of responsibility. Supervisors should contact our health and safety manager for advice and assistance. The standard accident investigation form should be used at all times.

Adapted from Goetsch, 2013

- Outlines of the various records that are both legally mandatory and for organisational purposes that will be kept.

Example 11.4: Example of an H&S record-keeping statement

CONSTRUCTION HEALTH AND SAFETY RECORD-KEEPING STATEMENT

CHRISTOPHER BUILDING CONTRACTORS

Christopher Building Contractors follows the current OHSA recordkeeping regulations and guidelines. Accurate and comprehensive records of all recordable injuries and illnesses are maintained and evaluated to identify trends and other causal data that can be used to reduce or prevent repeated occurrences. An annual summary of all recordable accidents and injuries is developed and reported in financial statements.

Adapted from Goetsch, 2013

- Details of any training to mitigate the exposures to specific hazards identified in the client's specifications.

Example 11.5: Example of an H&S training statement

CONSTRUCTION HEALTH AND SAFETY TRAINING STATEMENT

CHRISTOPHER BUILDING CONTRACTORS

All construction workers employed by Christopher Building Contractors are provided with regular construction health and safety training. New workers receive general training as part of their induction and site-specific training provided by their supervisor prior to starting work. Periodic updated training is provided annually and whenever a worker changes the type of work they do or moves to another site. Every week, supervisors provide toolbox talks at the beginning of a work shift on a relevant topic. Attendance at these training sessions is recorded.

Adapted from Goetsch, 2013

- Evidence of emergency procedures and medical and first aid response.

Example 11.6: Example of an H&S medical and first aid statement

CONSTRUCTION HEALTH AND SAFETY MEDICAL AND FIRST AID STATEMENT

CHRISTOPHER BUILDING CONTRACTORS

Prompt and competent medical services are available and provided to all our workers. First aid supplies are readily available on all our sites. We also have designated trained first-aiders to provide first aid and document all aid given in the first aid log. Where medical assistance beyond first aid is needed, this assistance is provided by calling the number of a designated medical practitioner. These numbers are posted conspicuously on all our sites. All our workers have full access to their own personal medical and exposure records.

Adapted from Goetsch, 2013

- Required risk assessment, method statements, safe working procedures and standards.

5. Pre-tender Phase Construction H&S Plan

The pre-tender phase construction H&S plan enables the inclusion of an adequate financial allowance in the tender or bid and is concerned specifically with the mitigation and response to the issues raised in the client's specification for the particular project. While it is not a regulatory requirement, a pre-tender plan is recommended as better practice. The following six items should be covered in this plan:

1. Description of the project
   - A detailed project description that describes, for example, the type of building or facility, the size or coverage of the building or facility, the height of the building or facility in terms of number of storeys, the nature of the structure and its construction and structural features, and programme details
   - Details of the client, designers and any other identified consultants
   - Extent and location of existing records and plans, such as historical maps, previous site investigation reports, meteorological reports, geotechnical reports, location and site plans, plans of existing structures, previous construction H&S files, and crime statistics, to name a few
2. Client considerations and management requirements
   - Structure of the organisation for the project and how construction H&S management on the project will be organised
   - Project- and site-specific H&S goals and how these are going to be monitored and reviewed

Example 11.7: Example of H&S goals on a project

> **ERECTION OF NEW BLOCK OF FLATS IN NEWLANDS**
>
> **PROJECT HEALTH AND SAFETY GOALS**
>
> **CHRISTOPHER BUILDING CONTRACTORS**
>
> - Zero fatal accidents on site during the project
> - No more than 5 injuries during the project
> - No more than 10 lost time accidents
> - Increase participation of workers in onsite training by 100%
> - These goals will be monitored and reported on during each weekly project progress meeting and will be reviewed when the projects reaches 50% completion.

- Details of any necessary permits and authorisation such as, for example, the permit to do construction work and how the process of obtaining them will be managed
- Emergency procedures for the construction site that include, for example, fire response and site evacuation with evacuation routes

Example 11.8: Example of an H&S emergency response statement

> **CONSTRUCTION HEALTH AND SAFETY EMERGENCY RESPONSE STATEMENT**
>
> **CHRISTOPHER BUILDING CONTRACTORS**
>
> **Fire:**
> All our workers at all our sites are trained in fire protection procedures and response. All sites are provided with properly serviced firefighting equipment. Emergency telephone numbers are conspicuously displayed.
>
> **Evacuation:**
> When it is necessary to evacuate a site or part of the site, a loud continuous siren is used to alert everyone on that site. The location of a 'safe area', where everyone must gather for a head count, is clearly and visibly signposted and identified to everyone on the site during induction. Supervisors are responsible for conducting a head count of workers directly reporting to them at the 'safe area'.

Adapted from Goetsch, 2013

- Details of complying with and enforcement of Covid-19 prescriptions on site
- Any site-specific H&S rules and other relevant restrictions that will affect contractors, suppliers and others, such as, for example, access to parts of the site that continue to be used by the client and security arrangements
- Details of any activities anticipated on or adjacent to the site during the contract duration
- Procedures to ensure proper liaison and communication between all the project stakeholders

3. Environmental restrictions and existing off-site and on-site risks
    (These could also be dealt with by separating out the health and the
    safety issues.)
    - Adjacent land use which might require special treatment during the project
      execution, such as working adjacent to a hospital that might involve
      restrictions on noise and deliveries
    - Traffic conditions and restrictions
    - Information about any possible ground instability and contamination, such
      as in parts of KwaZulu-Natal province during heavy rains
    - Location of existing services and utilities, such as water, electricity and gas
    - Condition of existing structures that might have to be worked in, including
      asbestos and fragile materials
4. Significant design and construction hazards
    - Details of any design assumptions for the structures to be erected on the
      site, if known, and control measures to be taken, such as suggestions for
      methods or sequence of erection or assembly
    - Arrangements for accommodating any ongoing design work, especially
      when the design is incomplete at the time of tender and details of how
      design changes will be managed
    - Response to any significant hazards identified during the design stage of
      the project
    - Details of any construction materials that might require special handling
      and precautions, such as tile panels that are required to be installed as a
      single unit
5. Construction H&S file
    - Information on how the compilation of the construction H&S file will
      be managed
    - An outline of the contents of the file and how such content will be collected
6. Budget
    - Since it is a requirement for the client to provide for the contractor to price,
      specifically for construction H&S management on the project in the tender,
      the contractors must include details of how the cost allowance has been
      calculated in response to the specific and significant hazards identified in
      the client's H&S specification.

6. Construction Phase H&S Plan

While the construction phase H&S plan will contain a substantial number of the
items included in the pre-tender version, it is likely that, after negotiation with the
client, there might be changes that have to be agreed to and incorporated before

any work commences on the project and construction site. The following five items should be covered in this plan:

1. Description of the project
   - A detailed project description that describes, for example, the type of building or facility, size or coverage of the building or facility, height of the building or facility in terms of number of storeys, the nature of the structure and its construction and structural features and programme details
   - Details of the client, designers, PC, contractors, plant suppliers, major material manufacturers and any other identified consultants
   - Extent and location of existing records and plans relevant to the construction phase of the project
2. Communication and management of the work
   - Final structure of the organisation for the project and details of how construction H&S management on the project will be organised
   - Agreed project- and site-specific H&S goals and how these are going to be monitored and reviewed by the contractor
   - Details of any necessary project and site specific permits and authorisation such as, for example, the permit to do construction work, and how the process of obtaining them will be managed together with timelines
   - Procedures to ensure proper liaison and communication between all the project stakeholders and exchange between them of project information that include any design changes
   - Selection, prequalification, appointment and control of contractors involved in the project
   - Establishment of legally mandatory worker forums, such as H&S representatives and committees and other worker engagement and consultative structures
   - Exchange of construction H&S information between contractors, including security arrangements, Covid-19 procedures, site induction, on-site training, regular toolbox talks, welfare facilities and first aid
   - Emergency procedures for the construction site that include, for example, fire response and site evacuation with evacuation routes
   - Any specific site H&S rules and other relevant restrictions that will affect contractors, suppliers and others, such as, for example, access to parts of the site that will continue to be used by the client and security arrangements
   - Details of any activities anticipated on or adjacent to the site during the contract duration
   - Reporting and investigation procedures of accidents, incidents and near misses
   - Development and approval of risk assessments and the implementation of mitigation interventions and method statements

3. Arrangements for controlling significant project and site hazards
(These could also be dealt with by separating out the health and the
safety issues.)
   - Details on how adjacent land use, which might require special treatment,
     will be managed during the project execution
   - Traffic routes, conditions, restrictions and segregation of vehicles and
     pedestrians on the site
   - Measures to deal with unstable existing structures
   - Location and protection of existing and temporary services and utilities,
     such as water, electricity and gas
   - Details of how fall prevention and lifting operations will be addressed on
     the construction site
   - Final details of any control measures responsive to the project design
     assumptions for the structures to be erected on the site, such as methods or
     sequence of erection or assembly and methods of handling
   - Arrangements for accommodating any ongoing design work and details of
     how design changes will be managed
   - Details of any construction materials that might require special handling,
     removal and precautions, such as asbestos and fragile materials
   - Provisions for the proper use, handling and storage of hazardous materials
   - Measures to be implemented to handle excessive noise and vibration
4. Construction H&S file
   - Information on how the compilation of the construction H&S file will
     be managed
   - An outline of the contents of the files and how these will be collected,
     including:
     - as-built drawings and plans
     - design criteria, such as safe and maximum loadings
     - potential residual hazards in the final structure
     - construction methods and materials used
     - plant and equipment maintenance requirements, intervals and procedures
     - operating and maintenance manuals for plant and equipment used in the
       completed structure
     - final location and nature of services and utilities
     - list with contact information of all consultants, contractors, suppliers and
       manufacturers involved in the project
5. Budget
   - The final negotiated and approved cost allowance to implement the
     measures to respond to the specific and significant hazards identified in the
     client's H&S specification.

Example 11.9: A template for a construction H&S plan

Topic Heading
Project introduction
Project management
Safety, health and environment standards set for the project
Information for contractors
Principal contractor's selection procedures for contractors, manufacturers and suppliers
Communications
Hazardous activities
Emergency plan or procedures
Reporting of incidents and other information
Welfare
Information and training for workers
Consultation arrangements
Site rules
Health and safety file
Monitoring arrangements
Project review

Adapted from Holt, 2005

7. Review Questions

- Do the Construction Regulations, 2014 require a formal written H&S plan?
- Why are both pre-tender and construction H&S plans considered better practice?
- What should be included in a construction H&S plan?
- What are the benefits of a construction H&S plan?

8. Review Exercises

- Compile a pre-tender construction H&S plan for a specialist tiling contractor bidding on the tiling contract in a 25 storey concrete-framed residence hotel, where the entrance lobby floor is to be finished using 1 200 x 1 200 m polished porcelain tiles
- Develop a construction H&S plan for a roofing contractor for this hotel, which has a tiled steeply pitched roof that has both hips and valleys.

Bibliography

Construction Industry Advisory Committee. 2003. *Designing for health and safety in construction.* London: HMSO.

Goetsch, DL. 2013. *Construction safety & health.* Upper Saddle River: Pearson Education, Inc.

Griffith, A & Howarth, T. 2014. *Construction health and safety management.* Upper Saddle River: Pearson Education, Inc.

Haupt, TC. 2019. *The state of construction health and safety in South Africa.* Lecture presented to final year engineering students and staff, Curtin University, Perth, Australia (10 September 2019).

Haupt, TC. 2019. *Challenges of construction in South Africa: Research opportunities to challenges of construction.* Lecture presented to staff, Massey University, Palmerston North, New Zealand (27 September 2019).

Haupt, TC. 2020. *Compliance or better practice: Health and safety in uncertain times.* Webinar delivered for University Extended Learning (13 October 2020).

Haupt, TC. 2021. *Management of safety, health and environment in South Africa: A handbook.* Newcastle Upon Tyne: Cambridge Scholar Publishing.

Health and Safety Commission. 2007. *Managing health and safety in construction: Construction (Design and Management) Regulations 2007: Approved code of practice.* Sudbury: Health and Safety Executive Books.

Health and Safety Executive. 2004. *Health and safety in construction.* Sudbury: Health and Safety Executive Books.

Hinze, J. 2006. *Construction safety.* Gainesville: Jimmie Hinze.

Holt, ASJ. 2005. *Principles of construction safety.* Osney Mead, Oxford: Blackwell Science Ltd.

Raliile, MT & Haupt, TC. 2019. Analysis of recent construction regulations changes and their impact on the quality of life of construction workers. *Proceedings of the 1st Association of Researchers in Construction Safety, Health and Well-being (ARCOSH) Conference,* Cape Town, South Africa (4 June 2019).

Raliile, M & Haupt, TC. 2020. The study on knowledge, attitudes and commitment of managers within construction firms towards recent construction health and safety legislation changes. *2nd Association of Researchers in Construction Safety, Health and Well-being Conference (ARCOSH),* Cape Town, South Africa (1-2 June 2020).

Raliile, M & Haupt, TC. 2020. Investigation of worker involvement in the implementation of health and safety policies on construction sites. Paper presented at the Joint CIB W099 & TG59 International Good Health, Wellbeing & Decent Work Conference, Glasgow, Scotland (10 September 2020).

Hazard Identification and Risk Assessment

Theo C Haupt

1. Introduction

The construction industry is a hazardous and dangerous one but not inherently so. Since the construction health and safety (H&S) legislative and regulatory framework in South Africa is a risk-based framework, there is a need throughout all phases of a construction project to identify hazards and manage the risk of exposure to them. Exposure to hazards on construction sites can be effectively managed to eliminate or reduce the consequences of exposure to these hazards. Where actions such as a construction activity can be deemed to have consequences, it is the degree of uncertainty under those circumstances that can be considered risk. Furthermore, every time a choice is purposely made, such as choosing a particular construction material to finish a façade or a certain construction technology to finish off a floor surface, risks are played off against each other, especially during a construction project and on site.

Risk is pervasive and can either be considered consciously or, as occurs most of the time, subconsciously. Risk can be perceived as a situation in which a decision is made, the consequences of which depend on the outcomes of future events having known probabilities.

2. Hazard Identification and Risk Assessment

Before the risk of exposure can be evaluated, it is important and necessary to identify the hazards that could lead to exposure to risks. These hazards refer to those arising from construction activities that require the use of machinery, equipment, tools, operating procedures, material handling and storage, processes and environmental working aspects of the construction site itself. The hazard identification process involves gathering information from previous experiences with executing similar construction activities under similar conditions, historic data, inspection reports, accident and incident reports and standards.

A hazard identification and risk assessment (HIRA) should be conducted to ensure the reduction or elimination of risk to those who will be involved in any stage of a construction process, project or installation, or during the maintenance of construction equipment on the construction site. The process does not need to be precise, but it should deal with known construction hazards, issues or risks where possible or reasonable.

Where this information is not readily available or the hazards may not be easily identified, other techniques may have to be used such as, for example, failure modes and effects analysis and job safety analysis (JSA). Logic diagrams are used in these techniques to understand why failures occurred. In the case of JSA, which was developed from work study practices, the hazardous activity is identified and the means to control the exposure to risk are applied. The activity is observed being executed and broken down into steps, with each step being thoroughly examined to determine the level of risk involved. Consultations with those who execute the activity are important in JSA. Obviously, while JSA is a thorough analysis, it is a time consuming and possibly expensive form of analysis.

Example 12.1: Example of the eight steps involved in a JSA

1	Identify the construction activity that will be evaluated
2	Observe the construction activity being performed
3	Break up the activity into sequential logical steps
4	Identify the hazard/s associated with each step
5	Consult with the construction workers involved
6	Decide on how the risk of exposure can be eliminated or reduced with appropriate interventions
7	Formally document the job site analysis process
8	Review the job site analysis as needed and update as applicable

The management oversight and risk tree is a simpler and more cost-effective technique that can be used.

A competent person is required to conduct the HIRA and a construction H&S specialist practitioner could be used to assist with this process. However, the HIRA process must commence during the initiation phase of the construction project and must be addressed during all the other construction project phases. For example, in the case of a major installation in the plant, the design related aspects could include, inter alia, concept design, in addition to detailed design, while evolving details, a schedule, method of fixing and specifications.

2.1 Developing the HIRA

The steps to be followed may vary slightly depending on the type of risk assessment used. Given the subjective nature of the process, it is advisable to put together a risk assessment team of at least three experienced persons to increase the likelihood of having a balance between those who are risk-averse and those who are risk-takers. However, the risks need to be determined from the initial concept or decision to proceed with the construction project, followed by an examination of the construction drawings, where they have been advanced to a reasonable stage and when the work is divided into the various activities afterwards. Thought must be given to how to limit exposure, to the method of procurement, to the project programme or schedule, to the construction processes to be followed and to the construction materials used.

Material safety data sheets (MSDSs) should be obtained from suppliers and manufacturers to determine both the short- or long-term health risks to construction workers and whether it is possible to use alternative construction materials and products where the risk of hazard exposure could have serious negative consequences. For example, have solvent-based paints been specified for use in a confined space in a structure where ventilation is unlikely to be optimum? In such a case, consideration should be given to whether a water-based paint could be used because, since personal protective equipment (PPE) may only be used as a last resort, the use of a solvent-based paint may mean that workers would have to be rotated in order to reduce the time they spend in that space, being exposed to the fumes of the solvents in the paint where no other safer or healthier option is available.

Previous knowledge is useful, especially where the construction work to be done is similar or identical to previously completed projects. However, it would be foolhardy to overlook the reality that no two construction projects, installations or maintenance operations are precisely the same. For example, while the placing of storm water channels in position on a civils road project invariably involves the same procedure, the environmental factors such as the nature of the ground and the need for shoring and support, topography, surrounding traffic on national roads or in farm and rural areas may alter the risks significantly. The safe installation of metal roof sheeting will depend on the height of the roof above ground level, the pitch of the roof and the type and length of the sheets. Workers need to be protected from falling and attachments to fall-arrest lines may be required to ensure their safety.

Since the HIRA may be required as part of a client's construction H&S specification and a contractor's construction H&S plan, and often is, it should be properly documented. Furthermore, the Occupational Health and Safety Inspectorate of the Department of Employment and Labour may require the responsible persons or department in the organisation to indicate what was done proactively to identify hazards and mitigate risks, particularly if an accident resulting in either fatalities or other injuries should occur.

A quantitative assessment process is recommended and risks should be graded from low to extreme. It must be remembered that the risk areas rated as medium to extreme must always be addressed. The HIRA needs to be made available to all parties involved in the construction project and should be kept on file in the construction site office for inspection, when necessary. Should it not be possible to eliminate or mitigate a hazard, the requisite insert should be made in the relevant documentation and, where appropriate, the financial implications should be shown.

2.2 Quantifying the risk

To make an informed decision with respect to deciding, among other things for example, whether a design, detail, method of fixing or specification is appropriate, the risk needs to be quantified. Risk can be explained as the likelihood that a specified undesired event will occur as a result of the realisation of a hazard. It is important to note that the consequences of exposure to every hazard are not the same in terms of **severity**, which is defined as the level of harm that an exposure would create. Therefore, this quantification allows for the risks to be compared and ranked. Based upon the principle of 'reasonably practicable', which includes cost and other goals, a decision can be made to amend or leave the design, or other relevant aspects, unchanged. Furthermore, a precise estimate is not required, as doing so would be too time consuming and often there is a lack of data. It is imperative that, as far as possible, access to good data is necessary to enable the proper evaluation and categorisation of the risk exposures. While there are many ways and techniques that could be used to rank the hazards and the risk of exposure to them, it is the decision that is made to mitigate the exposure that is critically important, rather than the method or technique used for quantification and ranking.

A simple approach involves a simple formula, namely:

> Degree of risk = severity or consequences or impact x probability or likelihood or frequency

The values used to estimate each element are not important as long as they can be used consistently to provide a reliable quantification of the risk of exposure. For example, the following four-point rating scale for severity and probability can be used (Example 12.2).

Example 12.2: Example of severity and probability rating scale

Severity/consequences/impact		Probability/likelihood/frequency	
Category	Value	Category	Value
Catastrophic Imminent danger exists, exposure to the hazard could result in death and/or illness/disease on a wide scale.	4	**Imminently Probable/Certain** Exposure is likely to occur immediately.	4
Critical Exposure to the hazard could cause serious illness/disease, major injury and/or property and equipment damage.	3	**Reasonably Probable** Exposure will probably occur in time.	3
Marginal Exposure to the hazard could cause illness/disease, injury or property and equipment damage but the results would not be expected to be serious.	2	**Remote** Exposure may occur in time.	2
Negligible Exposure to the hazard will not result in serious illness and/or injury beyond a minor first aid case.	1	**Extremely Remote/Rare** Exposure is unlikely to occur.	1

Therefore, exposure to a hazard that could result in a major injury but is extremely remote, when the values are inserted into the formula:

Degree of risk = severity or consequences or impact x probability or likelihood or frequency

it will produce a result of 3 (severity value of 3 x probability value of 1). Reversing the values from high to low will not affect the outcome or quantification.

It is advisable to quantify the risk exposure without the introduction of any interventions and then to redo the calculation with appropriate interventions in place until it is no longer reasonably practicable to reduce either the severity of the consequences of the exposure or the probability of the exposure or both the severity and probability. The hierarchy of H&S precedence proposed by the National Institute for Occupational Safety and Health (NIOSH) in the United States of America, and which is widely used, is recommended for use in the control of risks. The hierarchy, from most effective to least effective, is as follows:

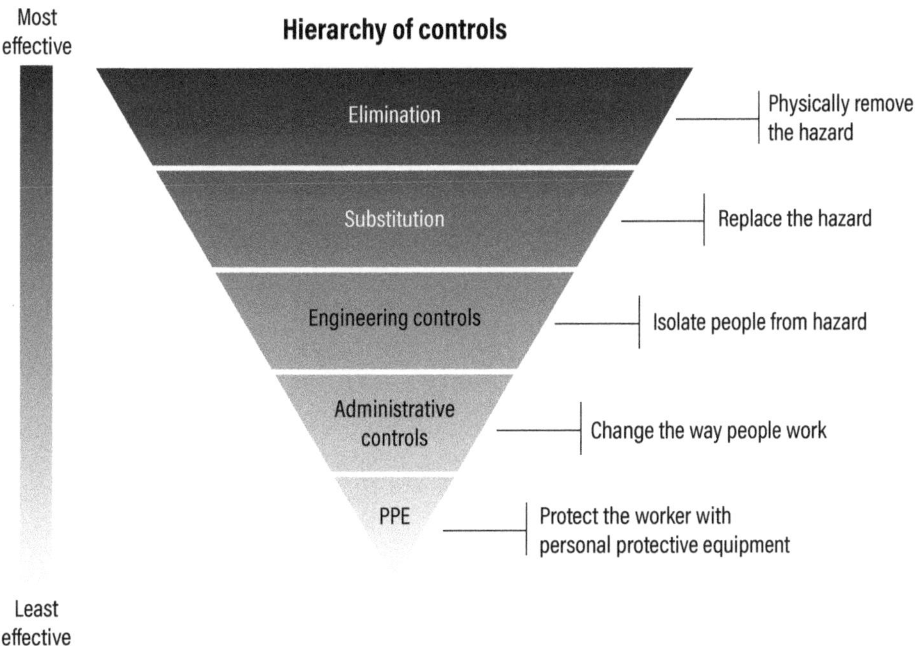

Figure 12.1: Hierarchy of H&S precedence (NIOSH)

- Elimination of a risk involves the use of alternatives or deciding not to attempt the work
- Substitution means where one material or substance can be replaced with another with less risk
- Use of barriers or engineering controls involves isolation, where the exposure to the hazard is removed from the construction worker, or segregation, where the worker is removed from the hazard
- Use of procedures or administrative controls, where exposure is limited through a process of rotation, for example safe working procedures
- Use of PPE, which should be used only when all other options have been exhausted, that is, as a measure of last resort.

It is also possible to quantify the risk of exposure to hazards by using a 3 x 3 matrix, as shown in Figure 12.2, where the risk can be determined by plotting the likelihood of the risk occurring (probability/frequency) on the vertical axis and the likely severity (consequences/impact) on the horizontal axis. The scores relative to probability and consequences/impact are as follows:

- Likely severity (consequences/impact), divided into three categories, namely high, medium and low:
  - high – fatality, major injury or illness causing long-term disability
  - medium – injury or illness causing short-term disability
  - low – other injury or illness
- Likelihood of occurrence (probability), divided into three categories, namely high, medium and low:
  - high – certain or near certain to occur
  - medium – reasonably likely to occur
  - low – very seldom or never occurs.

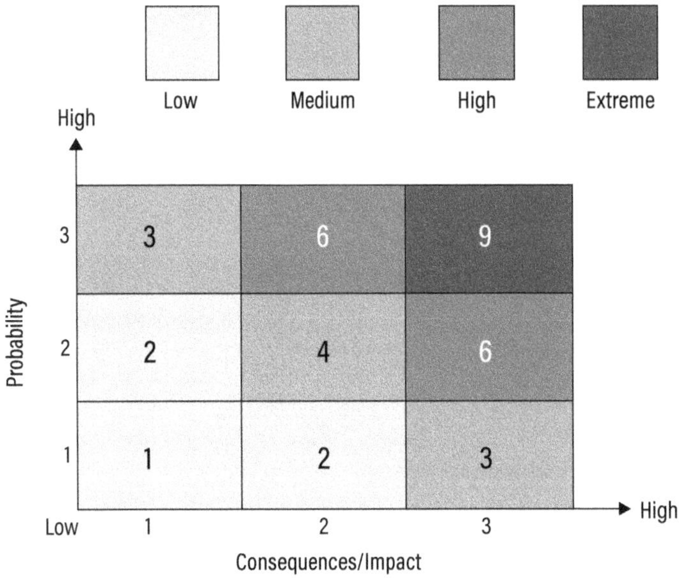

Figure 12.2: Probability/impact matrix (adapted from Burke, 2003)

Using this approach would obviously result in scores that are different from the previous approach (hierarchy of controls). A score of 9 using this approach would require elimination of the cause of the exposure to risk such as, for example, the selection of a flat roof in lieu of a steep pitched roof structure. A score of 6 would require elimination or, at least, substitution such as, for example, the use of a water-based paint in lieu of a solvent-based paint finish. A score of 4 would require elimination, substitution or, at least, containment by means of the use of a barricade such as, for example, screening of areas in which stripping of asbestos-containing materials would be required. A score of 3 would require elimination, substitution, containment or, at least, an engineering intervention, which would often also require that the construction work be done in accordance with a safe working or operating procedure (SWP/SOP), which would also entail the wearing of PPE such as, for example, a one metre high balustrade wall on a flat roof to prevent any falls. A score of 2 would require work in accordance with a SWP/SOP and entail the wearing of PPE such as, for example, erecting an external scaffold. A score of 1 would require no intervention.

The completed and approved HIRA needs to be communicated to the entire construction team, including the workers involved, and reviewed together with all identified controls. The final step in the HIRA is to evaluate and prioritise the risks, as all risks, their loss frequency and consequences are not the same.

Example 12.3 is an example of the pro forma risk assessment that a principal contractor could use.

Example 12.3: Risk assessment of a principal contractor

Project:							Document reference:					
Contractor:							Specialist discipline/activity:					
Assessor:				Signed:			Date:					

Activity/ element	Potential hazards	Population at risk	Risk rating P*C=R			Priority	Control measures specified	Hierarchy of risk control				
			P	C	R			E	S	EC	A	PPE

Legend:
P=Probability
C=Consequences
R=Risk
E=Elimination
S=Substitution
EC=Engineering control
A=Administrative control
PPE=Personal Protective Equipment

Sources of information:

3. Review Questions

- Why is a HIRA necessary?
- What is the most important aspect of a HIRA?
- Why is the method of quantification not a prescribed one?
- At what stage would PPE be an intervention in the HIRA process?
- Why would it be better for more than one person to conduct the HIRA for a project?

4. Review Exercises

- Using the following illustration and the pro forma risk assessment in Example 12.3, compile a HIRA for the construction activity in a renovation project inside a historical building

- The illustration shows the demolition phase of a construction project in progress using mechanical plant and equipment. Compile a HIRA for the demolition activities involved.

Bibliography

Burke, R. 2003. *Project management, planning and control techniques.* 4th edition. Chichester: John Wiley and Sons.

Griffith, A & Howarth, T. 2014. *Construction health and safety management.* Upper Saddle River: Pearson Education, Inc.

Haupt, TC. 2021. *Management of safety, health and environment in South Africa: A handbook.* Newcastle Upon Tyne: Cambridge Scholar Publishing.

Health and Safety Commission. 2007. *Managing health and safety in construction: Construction (Design and Management) Regulations 2007: Approved code of practice.* Sudbury: Health and Safety Executive Books.

Hinze, J. 2006. *Construction safety.* Gainesville: Jimmie Hinze.

Holt, ASJ. 2005. *Principles of construction safety.* Osney Mead, Oxford: Blackwell Science Ltd.

National Association of Home Builders (US). 2007. *Home builders' safety program.* Washington: Builderbooks.

CHAPTER 13

Permit to do Construction Work Application

Theo C Haupt

1. Introduction

Throughout the world when construction work is contemplated, various permits are required by responsible authorities to ensure that the projects can be monitored and inspected as needed and that the work complies with all applicable legislative and regulatory requirements. The permit to do construction work application is a formalised management control procedure that can be issued for any construction activity and usually applies to activities that are significantly hazardous. Provided the responsible authority executes its mandate with respect to construction work being executed within its jurisdiction, in the case of construction health and safety (H&S), the requirement by clients and the contractor to apply for specific permits prior to commencement of any work on the project site is a positive one. South Africa is not the only country where permits and notifications are required when it comes to construction H&S. For example, the United Kingdom has specific requirements in this regard.

2. Permit to do Construction Work

In terms of the Construction Regulations, 2014, any client who wishes to have construction work executed must, prior to the commencement of that work, apply in writing for a permit to do construction work. There are several conditions that must be present that will determine whether such a permit must be applied for. These are:
- the work to be performed must take more than 180 working days to complete
- the work to be performed must involve more than 1 800 normal working shifts of construction work by a worker
- the contract value of the project is equal to or exceeds R13 million or Construction Industry Development Board contractor grading level 6.

Therefore, not all construction projects need to have a permit to do construction work. The application, when required to be submitted, must be submitted to the relevant provincial director, accompanied by the following documents:

- baseline risk assessment (refer to Chapter 5)
- construction H&S specification (refer to Chapter 9)
- construction H&S plan (refer to Chapter 10).

An example of the format for the application is included as an annexure to the Construction Regulations, 2014 and is reproduced in Example 13.1 for ease of reference.

Example 13.1: Example of application for a permit to do construction work

<table>
<tr><td colspan="2">APPLICATION FOR A PERMIT TO DO CONSTRUCTION WORK
(in terms of Regulation 3(2) of Construction Regulations, 2014)</td></tr>
<tr><td colspan="2">This application must be submitted with the following three documents:
1. Health and safety specification
2. Health and safety plan
3. Baseline risk assessment</td></tr>
<tr><td colspan="2">1. Name, postal address and telephone numbers of the client:
Think Big Investment
P.O. Box 567
PRETORIA
0012
012 567 8912</td></tr>
<tr><td colspan="2">2. Details of the Agent:
    a. Mr Nobody, AB
    b. 223112 550209 2
    c. Xxxx1234
    d. 012 345 6789 (office) 076 102 7891 (mobile)
    e. Private Bag X100, PRETORIA 0001</td></tr>
<tr><td colspan="2">3. Name, postal address and telephone numbers of the appointed principal contractor:
ABC Contractors
P.O. Box 1234
PRETORIA
0012
012 123 4567</td></tr>
</table>

$\longrightarrow$

<table>
<tr><td>

4. Name, postal address and telephone numbers of the designer of the project:
21st Century Architects
P.O. Box 5678
DURBAN
4000
031 123 4567

</td></tr>
<tr><td>

5. Name, postal address and telephone numbers of the following persons:
   a. Construction Manager: Piet Pompies
   b. Construction Health and Safety Manager: John Doe
   c. Construction Health and Safety Officer: Jim Citizen

   All employed at: ABC Contractors
   P.O. Box 1234
   PRETORIA
   0012
   012 123 4567

</td></tr>
<tr><td>

6. Exact physical address of the construction and site office:
Cnr Eldorado and Soweto Roads
Thembisa

</td></tr>
<tr><td>

7. Nature of construction work:
Erection of a seven-storey residential block of apartments with two basement floors for parking constructed of a reinforced concrete framed structure with face brick infills between loadbearing columns and ring beams, aluminium windows glazed with safety glass, pitched tiled roof and skimmed rhino board ceilings etc.

</td></tr>
<tr><td>

8. Expected commencement date: 31 July 2022

</td></tr>
<tr><td>

9. Expected completion date: 31 August 2023

</td></tr>
<tr><td>

10. Estimated maximum number of persons on the construction site: 30

</td></tr>
<tr><td>

11. Planned number of contractors on site accountable to principal contractor: 8

</td></tr>
<tr><td>

12. Names of contractors appointed:
Red Roofing
White Painters
Pipe Plumbing and Drainage
Celestial Civils
Wet Plasterers
Square Tilers
ClearVu aluminium and glazing contractors
Bright Spark Electrical

</td></tr>
<tr><td>

13. Signature of Client/Client's Agent:

</td></tr>
<tr><td>

14. Signature of the Principal Contractor:

</td></tr>
</table>

$\rightarrow$

<table>
<tr><td colspan="3" align="center">FOR OFFICE ONLY</td></tr>
<tr><td align="center">Authorisation/Unique No.</td><td align="center">Labour Centre</td><td align="center">Official Approval Stamp</td></tr>
</table>

15. Date of application: 15 April 2022	
16. Submitted documents prescribed in Construction Regulations, 2014 5(4) (please tick ✓):	

CR 5(1)(a)	✓	CR 5(1)(b)	✓	CR 5(1); C-S)	✓

17. Result of the application (Please tick √):

Approved		Declined	✓

18. Reason for declining the application:
Submission of generic health and safety plan that is not responsive to the client's project and site-specific health and safety specification

19. Signature of the Supervisor:

----

20. Signature of the revoking Officer/Inspector:

The provincial director must issue a permit to do construction work, in writing, within 30 days of receiving the application, if satisfied that all the required information has been provided, verified and approved as adequate, and must assign a site-specific number for each construction site. It is the responsibility of the client to ensure that the principal contractor places a copy of the permit in the construction H&S file that is kept on site. The permit must be available to be inspected by an inspector, the client, the client, the authorised agent or a construction worker at all times.

Under no circumstances must any intended construction work be commenced before the permit to do construction work has been issued. The site-specific number assigned to the permit and particular project site may not be transferred to another site. In that case a new permit must be applied for.

3. Notification of Construction Work

Where the execution of any construction work that is not covered by the scope of the work outlined in the permit to do construction work applied for by the client is contemplated by a contractor, the contractor must notify the provincial director in writing no less than seven days before the commencement of that work. This need for notification by the contractor applies when the intended construction work involves excavations, working at height, demolition of an existing structure or elements of a structure or the use of explosives. The notification also requires information that will demonstrate to the provincial director that proper project management arrangements, including arrangements for construction H&S, have been made. Should it be necessary, this information will be used by the relevant authorities.

Where a contractor is appointed to construct a single-storey residential dwelling for a client who will live in the completed structure, the contractor must also notify the provincial director.

An example of the format for this notification is included as an annexure to the Construction Regulations, 2014 and is reproduced in Example 13.2 for ease of reference.

Example 13.2: Example of notification of construction work

<table>
<tr><td colspan="2" align="center">NOTIFICATION OF CONSTRUCTION WORK

(in terms of Regulation 4 of Construction Regulations, 2014)</td></tr>
<tr><td>1.</td><td>(a) Name and postal address of the appointed principal contractor:
ABC Contractors
P.O. Box 1234
PRETORIA
0012
012 123 4567</td></tr>
<tr><td></td><td>(b) Name and telephone number of principal contractor's contact person:
Gloria Poggenpoel – 082 123 4567</td></tr>
<tr><td>2.</td><td>Principal's compensation registration number: XYZ1234</td></tr>
<tr><td>3.</td><td>(a) Name and postal address of the client:
Think Big Investment
P.O. Box 567
PRETORIA
0012
012 567 8912</td></tr>
<tr><td></td><td>(b) Name and telephone number of client's contact person or agent:
Mr. Nobody, AB – 012 345 6789 (office) 076 102 7891 (mobile)</td></tr>
<tr><td>4.</td><td>(a) Name and postal address of the designer(s) for the project:
21st Century Architects
P.O. Box 5678
DURBAN
4000
031 123 4567</td></tr>
<tr><td></td><td>(b) Name and telephone number of client's contact person or agent:
Mr. Expert, CD – 012 789 1234 (office) 076 345 6789 (mobile)</td></tr>
<tr><td>5.</td><td>Name and telephone numbers of principal contractor's construction supervisor on site appointed in terms of regulation 8 (1): John Carpenter – 089 345 1267</td></tr>
<tr><td>6.</td><td>Name and telephone numbers of principal contractor's subordinate supervisors on site appointed in terms of regulation 8 (2): Peter Plummer – 082 345 9267; Gary Sparks – 076 231 4967</td></tr>
</table>

$\rightarrow$

<table>
<tr><td>7.</td><td>Exact physical address of the construction and/or site office:
Cnr Eldorado and Soweto Roads
Thembisa</td></tr>
<tr><td>8.</td><td>Nature of construction work:
Erection of a three-storey residential block of apartments with two basement floors for parking constructed of a reinforced concrete framed structure with face brick infills between loadbearing columns and ring beams, aluminium windows glazed with safety glass, pitched tiled roof and skimmed rhino board ceilings etc.</td></tr>
<tr><td>9.</td><td>Expected commencement date: 31 July 2022</td></tr>
<tr><td>10.</td><td>Expected completion date: 31 August 2023</td></tr>
<tr><td>11.</td><td>Estimated maximum number of persons on the construction site: 22
Male: 18 Female 4</td></tr>
<tr><td>12.</td><td>Planned number of contractors on site accountable to principal contractor: 6</td></tr>
<tr><td>13.</td><td>Names of contractors already selected:
Red Roofing
White Painters
Pipe Plumbing and Drainage
Wet Square Plasterers and Tilers
ClearVu aluminium and glazing contractors
Bright Spark Electrical</td></tr>
<tr><td>14.</td><td>Signature of the Principal Contractor:

Date: ____</td></tr>
<tr><td>15.</td><td>Signature of Client/Client's Agent (where applicable):

Date: ____</td></tr>
<tr><td>16.</td><td>Signature of Client:

Date: ____</td></tr>
<tr><td>17.</td><td>Signature of the revoking Officer/Inspector:

____</td></tr>
</table>

4. Review Questions

- Why do the Construction Regulations, 2014 require the client to apply for a permit to do construction work?
- Why would an application for a permit to do construction work be declined?
- Under what conditions must a principal contractor notify the provincial director of construction work intended to be executed?
- Why is the information requested in the notification necessary?

5. Review Exercises

- As a client who wants to erect a shopping mall, compile an application for a permit to do construction work.
- Compile a notification of construction work for the erection of a 20-bed double-story school dormitory.

Bibliography

Griffith, A & Howarth, T. 2014. *Construction health and safety management.* Harlow: Pearson Education, Inc.

Health and Safety Commission. 2007. *Managing health and safety in construction: Construction (Design and Management) Regulations 2007: Approved code of practice.* Sudbury: Health and Safety Executive Books.

Holt, ASJ. 2005. *Principles of construction safety.* Osney Mead, Oxford: Blackwell Science Ltd.

CHAPTER 14

Risk Mitigation Plan

Theo C Haupt

1. Introduction

Once completed, reviewed and finalised, the construction hazard identification and risk assessment (HIRA) must be incorporated into the construction project health and safety (H&S) management system to be applied on the project as a risk mitigation plan for its duration. The interventions identified and agreed upon in the HIRA must be implemented and reviewed for effectiveness, with the outcomes of the review noted for future reference and use in future HIRAs.

2. Management of Risk Mitigation Plan

The construction project-specific risk mitigation plan must clearly demonstrate how the significant risks identified and assessed during the HIRA process will be adequately resourced and managed during the execution of the project. Therefore, the anticipated exposures to hazards on the project should be mapped onto the project schedule or programme so that they can be effectively incorporated into the overall project management plan. It makes good business sense to appoint, in writing, someone with the appropriate level of authority who is suitably qualified in construction H&S to manage and oversee the implementation of the risk mitigation plan. The responsibilities of this appointee include, inter alia:

- ensuring that all construction work is done according to applicable standards of construction H&S, with minimum risk to all workers, equipment and materials
- monitoring compliance with all applicable construction H&S legislation, regulations, method statements, safe working procedures, construction project site rules and other instructions
- conducting regular formal H&S inspections while documenting all findings
- investigating all incidents, accidents and near misses to identify the causes and to develop interventions to mitigate future exposures and occurrences.

3. Content of the Risk Mitigation Plan

The construction risk mitigation plan must include specific basic procedures and clearly outlined minimum performance standards for construction activities. Apart from the specific provisions for mitigating the significant hazards identified in the HIRA, additional aspects of the construction project must form part of the mitigation plan and must include the safe working procedures and policies, namely:

- required construction H&S method statements
- procedures for control of secure and safe access to and egress from the construction site by construction workers and visitors
- controls for safe access to and egress from the site by motor vehicles and delivery vehicles
- procedures for the control and disposal of waste produced on the site by construction activities
- provision of, access to and maintenance of welfare facilities, such as ablutions and eating areas on site
- details of proper H&S signage to be displayed on the site
- erection of scaffolding and temporary support
- procedures for working at heights, confined spaces, excavations and demolitions
- procedures for working with various tools, plant and equipment
- procedures for proper use, care and maintenance of personal protective equipment and clothing
- procedures for monitoring, evaluation and treatment of construction health threats and exposures
- provision for fire and electrical safety
- site orientation and induction programme and procedures
- accident and incident prevention and reporting procedures
- procedures for the appointment of various construction H&S personnel
- programme for construction H&S education and training, including toolbox talks
- project emergency plan to deal with, for example, multiple injury accidents; fires that cannot be controlled and contained on the site; natural disasters, like floods, high winds and mud slides; and major structural collapses.

Example 14.1 is an example of a risk mitigation plan showing all the various elements of one.

Example 14.1: Example of a risk mitigation plan

Document History/Version Control

Issue Status	Issue Date	Prepared By	Reviewed By	Amended Section (s)	Review Due
Version 1.0	21 Jan 2022	TBA	A Other	N/A	21 Jan 2022
Version 2.0	1 Jun 2022	AN	Somebody	New numbering format	1 Jun 2022

Authorisation

Date	Title	Name
21 Jan 2022	Health and Safety Director	TBC

Introduction

Management of risk is an integral part of good business management. ABC Contractors has aligned its approach to risk management with the appropriate ISO standard and integrated the elements of the risk management process into its construction H&S management system.

Purpose

Outline the process for the management of risk for Project X in alignment with the appropriate ISO standard.

SCOPE

Applies to all ABC Contractors practices and processes on Project X.

References

Appropriate ISO standard.

Definitions

Risk: the effect of uncertainty on objectives.
Effect is a deviation from the expected (positive or negative). Objectives can have different aspects (eg financial, health and safety and environmental goals).

Related documents

XYZ
ABC

Risk management process

The risk management process is detailed in Figure 1 and each element is briefly discussed in the following subsections.

$\rightarrow$

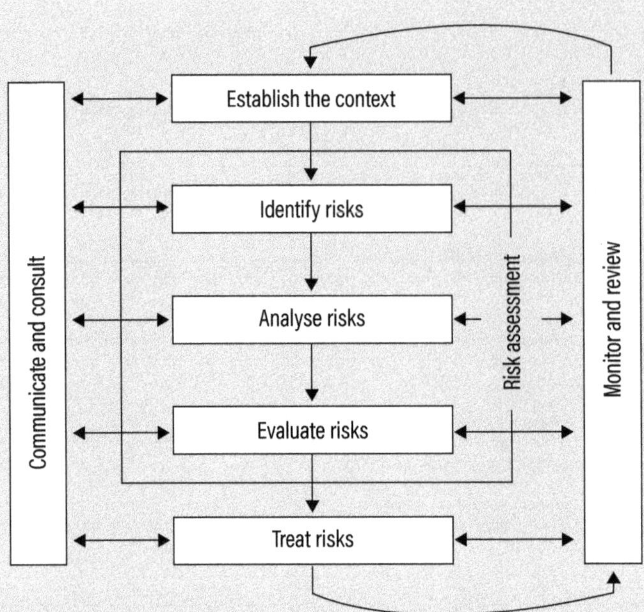

Figure 14.1: Risk management process

Communicate and consult

Communication and consultation should be undertaken with all relevant parties, as appropriate, at each stage of the risk management process and should concern the process as a whole.

Establish the context

The context in which the risk management process will take place should be established. This includes defining the external parameters, such as, for example, legal, regulatory, social, cultural and external project stakeholders, and internal parameters, such as, for example, construction health and safety policies, procedures and project objectives to be taken into account when managing risk, setting the scope and determining the criteria against which risk will be evaluated.

ABC Contractors has developed a risk assessment matrix to facilitate risk assessment, namely risk identification, analysis and evaluation (HIRA), including defining risk criteria.

IDENTIFY RISKS

Risk identification in the context of construction involves finding, recognising and describing risks arising from construction activities. This process involves the identification of risk sources and events, their causes and their potential consequences, such as, for example, prevent, degrade, delay or enhance the achievement of the project objectives.

ANALYSE RISKS

Risk analysis is a process undertaken to comprehend the nature of the risk and to determine the level of risk by considering the potential consequences or impacts of exposure and the likelihood or probability of them occurring. This analysis is achieved through the combination of semi-quantitative classes of consequence and likelihood defined in the risk assessment matrix. Consideration should be given to existing controls in the risk analysis.

EVALUATE RISKS

Risk evaluation is the process of comparing the results of a risk analysis, namely the estimated level of risk with risk criteria to determine whether the risk and/or its magnitude is acceptable or tolerable. This evaluation enables decisions to be made about the extent and nature of the treatments and interventions required and about priorities.

Risk criteria are defined in the risk assessment matrix.

TREAT OR MITIGATE RISKS

The requirement for risk treatment or mitigating interventions by implementing or modifying risk controls is determined by comparing the risk level against the risk criteria for treatment or mitigation. Risk treatment options are not necessarily mutually exclusive or appropriate in all circumstances.
Mitigating options can include:

- Avoiding the risk by deciding not to start or continue with the construction activity that gives rise to the risk
- Taking or increasing a risk in order to pursue an opportunity
- Removing the source of the risk
- Changing the likelihood or consequence of the risk
- Sharing the risk
- Retaining the risk by informed decision.

Selecting the most appropriate risk mitigation option involves balancing the costs and efforts of implementation against the benefits derived, with regard to legal, regulatory and other requirements. Where appropriate, consideration should be given to the hierarchy of control, namely elimination, substitutions, engineering, administration and personal protective equipment (PPE).

MONITOR AND REVIEW

It is necessary to review and monitor the effectiveness of all steps of the construction health and safety risk management process. This is important for continuous improvement of construction health and safety performance.

The monitoring and review processes should encompass all aspects of the risk management process for the purposes of:

- Ensuring that controls are effective and efficient in both design and operation
- Obtaining further information to improve risk assessment
- Analysing and learning lessons from events (including near misses), changes, trends, successes and failures
- Detecting changes in the external and internal context, including changes to risk criteria and the risk itself, which can require revision of risk treatments and priorities
- Identifying emerging risks.

RECORDING THE RISK MANAGEMENT PROCESS

The risk register is the primary record in construction health and safety risk documentation, with references to supporting documentation, where relevant.

4. Review Questions

- Who should be responsible for the implementation of the risk mitigation plan?
- What qualifications and experience should this person have?
- What should be included in a construction mitigation plan?
- How would it be possible to ensure that the correct mitigating interventions and resources are available at the correct time during the project?

5. Review Exercise

- Draft a job description with clearly stated responsibilities of the manager of the construction risk mitigation plan for the project depicted in the following photograph.

Bibliography

Griffith, A & Howarth, T. 2014. *Construction health and safety management.* Harlow: Pearson Education, Inc.

Haupt, TC. 2021. *Management of safety, health and environment in South Africa: A handbook.* Newcastle Upon Tyne: Cambridge Scholar Publishing.

Hinze, J. 2006. *Construction safety.* Gainesville: Jimmie Hinze.

Holt, ASJ. 2005. *Principles of construction safety.* Osney Mead, Oxford: Blackwell Science Ltd.

National Association of Home Builders. 2007. *Home builders' safety program.* Washington: BuilderBooks.

CHAPTER 15

Temporary Works

John J Smallwood

1. Introduction

Temporary works is a common term used globally when referring to work undertaken to support existing or future permanent works, the sides of excavations and to provide access.

The scope of temporary works includes:

- excavations
- support of existing or future structures
- access.

Earthworks include excavations.

Support of existing or future structures includes:

- falsework in the form of support work and formwork
- propping
- façade retention
- needling
- underpinning
- shoring of structures.

Access includes:

- fixed, mobile, special and suspended scaffolding
- temporary bridges.

This chapter refers to, but is not an extract of, the Construction Regulations, which can be referred to for more detail on certain types of temporary works.

2. Legislation and Regulations

The first issue to raise within the context of South Africa is the definition of temporary works in the Construction Regulations, namely that temporary works 'means any falsework, formwork, support work, scaffold, shoring or other temporary structure designed to provide support or means of access during construction work'. This a very narrow definition of temporary works, as opposed to what temporary works entails, hence the inclusion of the next chapter, Chapter 16: Ancillary Temporary Works.

The Construction Regulations, 2014 do refer to:

- structures
- temporary works
- excavation
- demolition
- scaffolding.

Part D of the National Building Regulations and Building Standards Act 103 of 1977 refers to demolition and Part G to excavations.

3. Temporary Works Issues

Planning is a key aspect of temporary works and should commence during Stage 1, project initiation and briefing because the client, construction project manager (CPM) or principal agent, construction health and safety agent (CHSA), designers and the quantity surveyor (QS) or cost engineer should consider how the permanent works will be executed. This should continue during Stage 2, concept and feasibility, and Stage 3, design development. Subsequent planning by the principal contractor (PC) and contractors during Stage 4, tender documentation and procurement, is critical and requires focus during the pre-contract and contract stages before commencing work on site. The project programme, project features, such as the site works, number of storeys, square metres per storey, structural frame, type of façade, type of party walls and finishes, and the bill of quantities provide the initial terms of reference. Site layout and method statements can then be evolved, which will provide information with respect to the nature and scope of temporary works.

Budgeting, which is an activity of planning, by the QS or cost engineer during Stage 4, tender documentation and procurement, should be appropriate and should not merely be based upon or extrapolated from previous projects as the project concerned may be substantially different from previous projects. Budgeting on the part of the PC and contractors should be based on detailed programmes, site layouts, method statements and scientific designs, where required.

Competence is a source of much debate and a **competent person** is defined in the Construction Regulations as a person who has the required knowledge, training, experience and, where applicable, qualifications specific to the work or task, and is familiar with the Occupational Health and Safety Act 85 of 1993, as amended (OHSA), and the applicable regulations and guidelines promulgated under the OHSA. The qualifications criterion is qualified by the requirement that, where appropriate, qualifications and training are registered in terms of the provisions of the National Qualifications Framework Act 67 of 2008, that those qualifications and that training must be regarded as the required qualifications and training 'specific to that work or task' should be noted.

Management by all the stakeholders is required throughout the relevant stages of the project. Given the nature of temporary works, competency relative thereto is critical. The need for, among others, linkage of temporary works to permanent works, activities that are part of processes to be linked and for multi-stakeholder contributions to be coordinated requires CPMs and construction managers who have received appropriate education and training. The diverse nature and scope of temporary works is such that a temporary works coordinator or manager should be appointed to manage and be responsible therefor. Furthermore, management must be committed and intimately involved in the related processes.

Supervision should be close throughout construction; however, the nature and related risk of many temporary works activities amplifies the need for close competent supervision. Examples include excavations, support of existing or future structures and access. Temporary works design should be assessed in terms of the process followed and the success thereof in the form of, among others, no failures. Temporary works execution can be assessed in terms of conformance to design, productivity relative to the process, conformance of the permanent structure in terms of quality standards, schedule performance and no failures.

There is a high level of risk with respect to many temporary works activities, which amplifies the need for competency, planning, adequate budgeting, scientific design and appropriate responses to the risks. Compliance with legislation, regulations and standards is a minimum endeavour, and better practice should be strived for.

4. Excavations

Historically, excavations have contributed to fatalities and other classes of injuries. For example, a trench collapsed during a new property development project near Klein Brak River, Mossel Bay in the Western Cape on 29 March 2005, which resulted in three worker fatalities. A trench also collapsed during construction of the Waterworld project in Randburg, Gauteng on 24 May 2005, which resulted in four worker fatalities.

Regulation 13, 'Excavation', of the Construction Regulations, 2014 addresses the subject in detail and schedules a range of requirements. Excavations must be supervised by a competent person (excavation supervisor), appointed in writing, and the stability of the ground should be determined before commencing excavation work. Fall and dislodgement of material must be prevented. Shoring and bracing are required unless the sides of the excavation are sloped to at least the maximum angle of repose relative to the horizontal plane, or an excavation takes place in stable material, in which case it must be confirmed by the competent person, appointed in writing. If there is uncertainty regarding stability, the decision of a professional engineer or a professional technologist competent with respect to excavations is required, which decision must be signed off by the aforementioned and the competent person.

The shoring must be scientifically designed and constructed in accordance with the design. Excavation spoil, materials and plant and equipment must not be positioned adjacent to the edges of excavations because of the resultant forces they exert. The stability of an adjoining building, structure or road must not be compromised by excavations. Safe access to and egress from excavations must be provided, and then not further than 6 m from where workers are working. Services must be identified prior to the commencement of excavations and interventions must be taken to ensure healthy and safe circumstances, which may include temporary support or even diversion. The competent person must inspect excavations daily, prior to the commencement of each shift, after every blasting operation, after an unexpected fall of ground, after damage to supports and after rain, and maintain a record thereof. Although the Construction Regulations state that excavations must be inspected daily, daily inspection is inadequate, regardless of the nature of the ground, because of their dynamic nature and the related hazards and risks. Where excavations are accessible to the public or adjacent to public roads or thoroughfares, they must be adequately protected by a barrier or fence of at least one metre high and provided with warning illuminants or clearly visible boundary indicators at night or when visibility is poor. People entering excavations must follow the precautionary measures stipulated for confined spaces as per the General Safety Regulations, 2003. When using explosives, a person competent in the use of explosives must be appointed and a method statement must be developed for the blasting process. Warning signs must be erected adjacent to excavations in which workers are working or in which persons are conducting inspections. This is necessary to prevent people from being buried beneath backfill material or being struck by mechanical equipment, such as an excavator bucket.

5. Support of Existing or Future Structures

Falsework is required to provide temporary support, primarily, during the construction of future reinforced concrete structures. Shoring, in turn, is required to provide support of existing structures. Both falsework and shoring must be able to support and withstand a range of loads and entail work at heights, which can result in the exposure of people to hazards and risks.

5.1 Falsework

Falsework (support work and formwork) failures have featured globally, and South Africa is no exception. Consequently, the Construction Regulations, 2014 are comprehensive with respect to this and related aspects of temporary works in the form of Regulation 12, 'Temporary works'.

Key requirements include:

- appointment of a temporary works designer to design, inspect and approve the falsework erected before use
- designs should be related to the structural design drawings and, where any uncertainty exists, the structural designer should be consulted
- design drawings must be available on site
- inclusion of construction sequences and method statements in temporary works drawings
- supervision of all activities by a competent person, appointed in writing
- a fall protection plan
- evaluation of the workers' medical fitness
- provision of fall prevention and fall arrest equipment
- workers must be trained and instructed
- equipment must be inspected before use
- adequate foundation conditions, including founding of supports
- adequate erection, support, bracing and maintenance to enable support in terms of anticipated vertical and lateral loads
- safe access for work above foundation level
- appropriate signage
- inspection by a competent person before, during and after the placement of concrete, after inclement weather or any other imposed load, and at least on a daily basis, until the temporary works structures are removed
- securing of soffits
- application of adequate release agents
- approval before casting of concrete
- achievement of sufficient concrete strength before removal of falsework and formwork
- approved back propping.

It should be noted that within the construction industry there are reservations about one person fulfilling the three-function appointment in terms of a temporary works designer to design, inspect and approve the falsework in the form of Regulation 12(1). These reservations were addressed by the Guidelines to the Construction Regulations, 2017, which state that the temporary works designer could be one person or different persons whose task would be to design, inspect and/or approve. However, from a construction management and supervision perspective, there are further issues, among others, the steel reinforcing installation, cover (spacing) between the outer reinforcing steel and the face of the concrete element, appropriateness of course aggregate size, the installation of electrical and other services, sleeves for the transmission of services and void formers for future openings, which do not fall within the scope of service of a temporary works designer.

5.2 Shoring

Shoring is a form of temporary lateral or vertical support for an unstable element of a building or structure. Raking and flying shoring provides lateral support, while dead shoring provides vertical support. Shoring may be required because of the demolition of an adjacent building, bulging or cracks in walls, unequal settlement of foundations, lateral support to walls as a result of adjacent excavations and temporary support of a historical façade to be attached to and incorporated into a new building. Dead shoring may be required in conjunction with underpinning when rebuilding the lower part of a structure, replacing deteriorated walls or sections thereof and creating large openings in existing walls. The duration of the shoring depends on the application and may be required for a week or even a year or more. Consequently, shoring should be robust and durable, in addition to be able to support the loads. Shoring is generally constructed from timber, steel sections or steel tubing.

Key requirements relative to all three types of shoring include:

- the appointment of a temporary works designer to design, inspect and approve the shoring upon completion
- designs should be related to the building or structure they are intended for and not be taken from a prior application
- design drawings must be available on site
- the design should indicate the construction sequence and key interventions, ie a design and construction method statement
- equipment must be inspected before use
- use of structural grade timber
- adequate foundation conditions, including founding of platforms or sole plates
- attachment of rakers to platforms and dead shores to soleplates
- inclination of rakers (raking shores) and struts (flying shores)
- rakers and walls (flying shores), and shores and struts (flying shores) should meet at floor level
- secure attachment of rakers, dead shores and struts to the structure

- spacing of shores
- horizontal and cross bracing between shores
- securing of wedges
- appropriate signage
- a fall protection plan
- evaluation of the workers' medical fitness
- provision of fall prevention and fall arrest equipment
- supervision of all related activities by a competent person appointed in writing
- erection and dismantling by competent workers
- safe access for work above foundation level
- inspection by the competent person after inclement weather, especially in the case of raking or dead shoring, and on a weekly basis until the shoring is removed
- maintenance, where necessary
- approved removal upon the building or structure having required the requisite strength.

Considerations include the duration of the shoring, site layout, the passage of vehicles and the position of future services.

6. Access

Access includes a range of scaffolding, suspended platforms mast climbing work platforms (MCWPs) and temporary bridges. Although this section only addresses fixed scaffolding in terms of scaffolding, many key requirements and considerations apply to the other types of scaffolding.

6.1 Fixed scaffolding

The South African construction industry has experienced numerous scaffold-related incidents and accidents that have resulted in injuries and fatalities. A notable accident is the Investec scaffold collapse on 26 August 1997. Three workers died and 16 were seriously injured when six floors of external scaffolding collapsed.

Regulation 16, 'Scaffolding', of the Construction Regulations, 2014 refers to and relies on South African National Standards (SANS) 10085-1:2004. 'The design, erection, use and inspection of access scaffolding', Part 1: 'Steel access scaffolding' for guidance.

The key requirements of a range of scaffolding-related aspects include:
- scientific design
- solid foundation
- appropriate soleplates
- baseplates for vertical components

- appropriate vertical and horizontal components and bracing
- attachment to and distance from the permanent structure
- adequate platforms in terms of width and envisaged loads
- guardrails
- toe boards
- screening
- access and access covers
- appropriate signage
- overhead catch platforms, where there is a danger of falling materials or equipment
- supervision of erection, alteration and dismantling by a competent person, appointed in writing
- a fall protection plan
- evaluation of the erectors' and supervisor's medical fitness
- provision of fall prevention and fall arrest equipment
- inspection by a competent person, appointed in writing before use, at designated intervals and after rain, and the maintaining of a record thereof
- safe loading diagrams to mitigate overloading the platforms
- appropriate related signage
- maintenance.

Considerations include:
- envisaged loads in terms of people, equipment and materials
- protection of the public from components, where scaffolding encroaches on a public thoroughfare
- installation of alarms, when scaffolding encroaches on a public thoroughfare, to prevent public access to upper levels, and screening to prevent falling materials and equipment
- mitigate against workers experiencing wind chill
- mitigate against visual pollution, provided such screening does not become a hazard during high winds.

6.2 Suspended platforms

Suspended platforms are addressed by Regulation 17, 'Suspended platforms', of the Construction Regulations, 2014, which constitutes the most comprehensive sub-regulation in the Construction Regulations. The reason for this is likely because of the frequency of suspended platform-related accidents and fatalities, among others, the suspended platform collapse in Hillbrow, Johannesburg in Gauteng on 6 February 2001, which resulted in six fatalities.

Key requirements include:

- a competent person appointed, in writing, to supervise all suspended platform work operations
- competent erectors, operators and inspectors
- issue of a certificate of system design (CSD) must be issued by a professional engineer, a certificated engineer or a professional technologist
- development of an operational compliance plan (OCP). The OCP must be developed by a competent person, based on the CSD, and include:
  - competency of erectors, operators and inspectors
  - operational design calculations (ODCs)
  - performance test results
  - sketches indicating the completed system and the operational loading capacity
  - procedures for and records of inspections and maintenance work having been carried out.

A copy of the CSD, the ODCs, sketches and test results, and the intended type of work the system will be used for, must be submitted to the provincial director of the Occupational Health and Safety Inspectorate, Department of Employment and Labour, before commencement of the use of the system for every project. Outriggers must have a safety factor of at least four relative to the intended load and the suspension points must be provided with stop devices at the outer ends to prevent the displacement of ropes.

Further key requirements include:

- capacity of the building or structure supporting the outriggers
- separate connection of the suspension wire rope and the safety wire rope to the outrigger
- wearing of body harnesses as a fall prevention device, which must be attached to the suspended platform
- mechanical prevention of uncontrolled movement of the working platform
- verticality of the rope connections between the outriggers and the working platform.

The suspended platform must be close to the structure as possible; have anchorage points to attach the lanyards of safety harnesses and indicate the maximum mass load, maximum number of persons and total mass load, including load and persons.

The contractor must ensure that the whole installation and all working parts of a suspended platform are thoroughly examined by a competent person in accordance with the manufacturer's specification. The contractor must also ensure that the whole installation is subjected to a performance test by a competent person, appointed in writing, who must determine the serviceability of the structures, ropes, machinery and safety devices before they are used and every time suspended platforms are erected, and the performance test of the whole installation of the suspended platform to be subjected to a load equal to that prescribed by the manufacturer or, in the absence of such load, to a load of 110 per cent of the rated mass load, at intervals not exceeding 12 months. Furthermore, the contractor must ensure that every hoisting rope, hook or other load-attaching device that forms part of the suspended platform is thoroughly examined by the competent person before they are used every time they are assembled and, in cases of continuous use, at intervals not exceeding three months. The suspended platform supervisor or the suspended platform inspector must conduct daily inspections to establish functionality, fitness for use, compliance with the requirements in terms of load restrictions and operation. Inspection and performance test records must be always available on site.

All workers operating or using a suspended platform must be medically fit to work safely in a fall risk position and be competent in terms of working safely thereon, and be trained on a range of issues, including procedures. The outriggers must only be moved by persons trained to do so and under the supervision of the competent person in accordance with the limitations of the OCP, followed by a documented inspection by the supervisor. Suspended platforms must be isolated after use at the end of shifts.

6.3 Mast climbing work platforms

Mast climbing work platforms (MCWPs) are used to perform work at height and are either single mast, in the case of small platforms, or larger multiple mast, in the case of platforms for full elevations. They provide both access and a platform, and can accommodate people, materials and equipment.

Key requirements include:

- an adequate structure to which the MCWP will be attached
- supervision of the design and erection by competent persons, appointed in writing
- a fall protection plan
- evaluation of the riggers' and workers' medical fitness
- provision of fall prevention and fall arrest equipment
- competent riggers
- competent operators
- a level, firm and stable foundation for the mast(s) and base
- use of the base outriggers, when necessary
- suitable guardrails and toe boards to the platform
- intermittent securing of the mast to the permanent structure in accordance with manufacturer's guidelines
- a means to prevent the platform falling, with overspeed
- adequate indication of the safe working load
- overload warning device

- fencing off of the area around the base to prevent people being struck by the platform
- tests, inspections and maintenance.

Considerations include the:
- required working width
- interaction between wind, the permanent structure and MCWP
- number of people to be supported
- the mass of materials and equipment to be supported.

6.4 Temporary bridges

Temporary bridges may be required in certain projects. Such bridges may span railways, roads or watercourses. The principles, key requirements and considerations for temporary bridges are similar to those for falsework in the form of support work and formwork. However, the implications of failure are more profound because temporary bridges invariably span public thoroughfares. The collapse of the M1 motorway temporary pedestrian bridge during afternoon traffic on the 14 October 2015 resulted in 120 tonnes of steel crashing on to the highway, and the loss of two lives and 19 injured people, some severely. In essence a temporary works designer was not appointed to design, inspect or erect the bridge. The PC used general arrangement drawings provided by the supplier of the components, which were only meant to determine the number of components needed and to compute the price to hire the components. These drawings were never intended to constitute the formal design for the erection of the temporary structure.

7. Review Questions

- Are the Construction Regulations adequate in terms of addressing the range of temporary works undertaken in construction?
- Is the current level of construction temporary works education, training and continuing professional development in South Africa optimum?
- Does the Occupational Health and Safety Inspectorate of the Department of Employment and Labour interrogate temporary works design processes to the necessary extent?

8. Review Exercises

- Compile a 750-word article titled *Assuring healthy and safe temporary works in general.*
- Compile a flow progress diagram illustrating the common interventions required to assure H&S temporary works in general.

Bibliography

Cresswell, R. 1997. The men who walk the planks. *Sunday Tribune*, 31 August 1997: 12.

Molwedi, P. 2001. I saw them fall like leaves. *The Star*, 7 February 2001: 1.

Myer, D. 2005. Trench deaths – can they be avoided? *National Safety & Occupational Hygiene*, May/June: 5–7.

Slabbert, A. 2019. Damning report into M1 highway bridge collapse that killed two. *City Press*, 2 December 2019, https://www.news24.com/citypress/business/damning-report-into-m1-highway-bridge-collapse-that-killed-two-20191202.

CHAPTER 16

Ancillary Temporary Works

John J Smallwood

1. Introduction

Within the context of this chapter, ancillary temporary works refers to work undertaken to support or rig an item of plant or equipment, or earthworks other than excavations, to accommodate and protect or to facilitate construction.

The scope of ancillary temporary works includes:

- the categories of site accommodation
- protection
- earthworks
- plant and equipment support
- access
- site manufacturing works
- groundworks for plant operation.

Site accommodation includes:

- site fencing or hoarding
- guardhouses
- site offices
- materials' stores
- welfare facilities
- services
- signage
- living accommodation
- fire prevention and protection.

Protection includes:

- catch platforms
- edge protection
- barricades
- signage.

Earthworks include:

- temporary slopes
- stockpiles
- temporary roads.

Plant and equipment support includes:

- bases
- supports
- anchors and ties for tower cranes, material and/or personnel hoists
- mast climbing work platforms (MCWPs)
- temporary electrical installations
- cantilever loading platforms
- rubble chutes.

Site manufacturing works include bulk mixing plants.

Groundworks for plant operation include suitable platforms for plant erection such as mobile cranes and piling rigs.

This chapter refers to, but is not an extract of, the Construction Regulations, which can be referred to for more detail on certain types of ancillary temporary works.

2. Legislation and Regulations

The Construction Regulations, 2014 do refer to:

- structures
- material hoist
- bulk mixing plant (batch plant)
- cranes
- construction vehicles and mobile plant
- electrical installations and machinery on construction sites
- use and temporary storage of flammable liquids on construction sites
- stacking and storage on construction sites
- construction employee facilities.

Pard F of the National Building Regulations and Building Standards Act 103 of 1977 refers to site operations, such as protection of the public, builder's sheds and sanitary facilities.

3. Ancillary Temporary Works Issues

Although ancillary temporary works issues are similar to, if not alike, the temporary works issues which were addressed in detail in Chapter 15, the importance of planning, budgeting, competence, management, the appointment of a temporary works coordinator or manager, supervision, the assessment of temporary works design and execution, the level of related risk, and compliance should be noted.

4. Site Accommodation

Site accommodation facilitates and is an integral part of the construction process and its activities due to its impact on the environment, health and safety (H&S), productivity, schedule and worker satisfaction. Furthermore, site accommodation impacts on the image of the related contractor and the construction industry.

4.1 Site fencing or hoarding

Site fencing or hoarding is important because of a range of issues, namely:
- legislation and regulations
- containment of site activities
- the health and safety and protection of the public
- security
- access control.

Part F1, 'Protection of the public', (1) to (3) of the National Building Regulations and Building Standards Act requires the erection of a fence, hoarding or barricade.

Key requirements include access for entry and exit of personnel, plant and vehicles, durable, theft proof, vandalism proof and are reusable. Personnel and vehicular access will require entrance control in the form of booms, gates or roller concertina closures.

Considerations include branding, external public relations and projected image. In terms of external public relations, the provision of viewing panels, in the case of solid hoarding, enables the public to view the project. Efforts to integrate the construction site into a neighbourhood could include the suspension of plants in planters from the hoarding. Figure 16.1 illustrates a protected public walkway to a project in Mouille Point, Cape Town, constructed from containers. In addition to failsafe protection of the public, the corporate colours and branding are notable.

Figure 16.1: Protected public walkway to a project, Mouille Point, Cape Town (Smallwood, 2015)

4.2 Guardhouse(s)

A guardhouse or guardhouses may be required to ensure protected access control to site by personnel, plant and vehicles. Key requirements for guardhouses include weatherproofing, durability, provision of services, functionality and security. Considerations include branding, internal and external public relations, and projected image.

4.3 Site offices

The scope of site offices depends on the size of a project, however, regardless of their size, they should facilitate administration, meetings and site visits by various project stakeholders. Key requirements include weatherproofing, durability, noise proofing, provision of services, illumination, ventilation and temperature control, functionality and security. Considerations include branding, internal and external public relations and projected image.

4.4 Materials stores

Key requirements for materials stores include durability, weatherproofing, illumination, functionality, security, proximity to the building or structure to be constructed, accessibility in terms of deliveries, compliance with legislation and regulations, supervision by a competent person, appointed in writing and maintenance in the form of housekeeping. Considerations include the level of security required relative to the respective materials to be stored, the mass of materials, potentially hazardous nature of materials, subcontractor requirements, branding and projected image.

4.5 Welfare facilities

The Facilities Regulations, 2004 are comprehensive and are referred to in Regulation 30, 'Construction employees' facilities', of the Construction Regulations, 2014, which are prescriptive. Part F11, 'Sanitary facilities', of the National Building Regulations and Building Standards Act refers to 'sanitary facilities' and compliance with the South African National Standards (SANS) 10400-F. 'Site operations'.

Key requirements include:

- shower facilities, based upon consultation with the employees or their representatives, or at least one shower for every 15 workers
- at least one sanitary facility for each gender and for every 30 workers
- changing facilities for each gender, including the provision of seating
- sheltered eating areas.

Further requirements include:

- illumination
- ventilation
- ancillary facilities in the form of wash hand basins
- the provision of soap, paper towel dispensers or hot air hand driers, related wastebins and hand sanitisers because of the Covid-19 pandemic and, for that matter, exposure to hazardous chemical substances.

Considerations include respect for people, internal and external public relations and image as the construction industry in developing countries is not renowned for welfare facilities. Consequently, the frequency of inspection, maintenance and cleaning of such facilities is critical. Furthermore, the Construction Regulations do not refer to lockers, which begs the question: Where do workers store their personal effects while in the workplace?

4.6 First aid facilities/occupational health screening

Regulation 3 of the General Safety Regulations, [Updated to 18 July 2003] 'First aid, emergency equipment and procedures', deals with first aid requirements; however, the regulation does not state that a first aid room is required.

Key requirements include:

- a first aid box when more than five people are employed in a workplace
- a qualified first-aider when more than 10 people, and for every group of up to 50 people, are employed in a workplace
- first aid box sign indicating the location and the name of the first-aider
- an eye wash fountain, if the potential of injury to the eye through contact with a biological or chemical substance
- a fast-reacting deluge shower with clean water, if there is a potential hazard of injury to or from absorption through the skin as a result of sudden contact with a large amount of a toxic, corrosive, high risk or similar hazardous substance.

These requirements constitute minimum requirements and do not reflect better practice. Ideally, a first-aider, or more than one first-aider, would have a dedicated first aid facility, which could be used for occupational screening and wellness counselling, in addition to rendering first aid. A further consideration is that an employee being treated may require a bed to lie on. Therefore, an appropriately equipped first aid room should be provided on projects when a first aider is required.

4.7 Services

The required services are dependent upon the degree of permanence, nature and scope of the site accommodation and plant and equipment.

Key requirements generally include electricity, single phase and/or three phase; standby generators; information and communications technology; and water and sewage, in the case of flush toilets, shower facilities and wash hand basins.

Considerations include respect for people, internal and external public relations and image.

4.8 Signage

Although the OHSA, the Construction Regulations and the National Building Regulations and Building Standards Act schedule minimum H&S-related signage requirements, a need exists for signage related to project requirements, such as the project notice board, which records the information of the primary stakeholders. Further signage includes that related to information and branding.

The seven H&S-related categories of signage include:

1. mandatory, for example wearing of hard hats
2. prohibition, for example no smoking
3. warning, for example uneven surfaces
4. danger, for example the potential for electrical shock
5. fire safety, for example the position of fire extinguisher
6. emergency information, for example emergency assembly point
7. restriction, for example speed limit.

Key requirements include:

- relevance
- appropriate position
- plumb/level
- logical, ie no upside down no smoking signs
- securely attached or fixed
- regular cleaning
- addressed during dedicated toolbox talks.

4.9 Protection

Protection includes catch platforms, edge protection and barricades.

Catch platforms, which are generally attached to the sides of buildings or structures under construction, are intended to protect both workers on site and the public using adjacent public thoroughfares, such as sidewalks and roads, from falling equipment, materials and tools.

Edge protection is primarily related to workers working at elevated heights, although it may, on occasion, be necessary to prevent members of the public from falls because of the proximity of public thoroughfares to construction activities.

Barricades are invariably related to open excavations that affect both workers and members of the public, depending on the locations of the excavations.

Key requirements include appropriate height; the ability to resist lateral loads and the transmission of materials or tools; must be discernible at night; easy to erect and dismantle; and the potential for reuse.

Considerations include internal and external public relations and projected image, which is amplified by the degree to which catch platforms, edge protection and barricades are visible to the public.

4.10 Living accommodation

Regulation 30(2), 'Construction employees' facilities', of the Construction Regulations, 2014 states that contractors must provide reasonable and suitable living accommodation for workers working on construction sites far from their homes, where adequate transportation between the site and their homes or where other suitable living accommodation is not available. The terms 'reasonable' and 'suitable' are not defined, neither is the term 'far'.

Key requirements include:

- separate accommodation for workers of different genders
- a separate bed for each worker
- separate lockers for personal belongings
- potable drinking water
- adequate sanitary and washing facilities
- adequate ventilation
- adequate protection against cold, heat, damp, noise, fire, disease-carrying animals and insects
- canteens
- rest and recreation facilities.

Considerations include:

- the duration that the living accommodation is required
- the lifestyles of the accommodated workers
- leisure time facilities, such as television rooms and games rooms
- the provision of health care facilities.

4.11 Fire prevention and protection

Construction sites are susceptible to fires and a fire risk assessment is critical during the pre-tender stage when preparing the site layout because of the need to incorporate the necessary provisions in terms of fire prevention and to make adequate financial allowance for fire protection. The potential sources of ignition, fuel and oxygen should be identified. The potential contribution of temporary electrical installations and the storage of flammable goods, in terms of ignition and fuel, amplify the need for fire prevention and precautions.

Key requirements include:

- limiting the spread of fire through compartmentation and separation
- fire detection
- raising the alarm
- fire exits and escape routes
- emergency planning
- firefighting equipment.

Compartmentation can be achieved by programming the permanent works in such a manner that sections of floor areas can be compartmentalised by the enclosing of the structure and by the hanging of doors, especially fire doors. Generally, extinguishers are the preferred form of firefighting equipment, which should be appropriate in terms of the potential fire:

- water, foam or multipurpose dry powder extinguisher for cardboard, cloth and timber fires
- dry powder or foam extinguisher for flammable liquids
- carbon dioxide or dry powder extinguisher for electrical-related fires.

5. Earthworks

Earthworks other than excavations include temporary slopes, stockpiles and temporary roads.

5.1 Temporary slope protection

The objective of temporary slope protection is to reduce the amount of water flowing onto the surface of the slope because, if the slope becomes saturated with water, its strength and stability are likely to decrease. Slope failures result from the movement of material downslope as a result of the pull of gravity and are invariably triggered by the loss of strength resulting from the slope's saturation with water. There are environmental, H&S, productivity and rework implications should a slope fail.

The amount of water running down the slope should be minimised by a surface water diversion at the top of the slope and the slope should be covered with plastic sheets or a geotechnical membrane, restrained with earth filled sandbags at the top of the slope and at intermediate points, in rows, between the top and the bottom of the slope. The toe of the slope should not be disturbed and neither should it become saturated or else slope failure will be induced.

5.2 Stockpiles

Stockpiles may be required as a construction activity or they may be related to quarrying.

Key requirements include a level, firm, stable and drained foundation that is not likely to be threatened by water courses.

Considerations include:

- the likely volume of the stockpile
- the stability and angle of repose of the stockpiled material
- area available, including access by mobile plant to dump and load out
- potential hazards, such as power lines in the proximity of the stockpile
- high winds
- the potential environmental impact.

5.3 Temporary roads

Projects may be of such a nature that temporary roads are required, as in the case of multi-building projects, challenging soil conditions, haul roads, in the case of infrastructure projects, and detours, in the case of civil engineering projects. Regulation 23, 'Construction vehicles and mobile plant', of the Construction Regulations, 2014 provides partial information with respect to roads by referring to the separation pedestrians (workers) and vehicles, number and size of traffic routes and the provision of signage.

Key requirements include:

- appropriate slope or camber
- drainage to convey water runoff
- road signage, such as speed restrictions
- separation of people from plant and vehicles
- adequate provision for turning
- reinstatement at the end of the project.

Considerations include:

- sensitive areas in terms of flora and fauna
- expected axial loads
- one-way or two-way traffic
- potential of erosion
- type of material versus the slope or camber.

6. Plant and Equipment

Various types of plant and equipment are required to move people, materials and other plant and equipment, which can result in hazards and risks. Furthermore, such plant and equipment affect performance relative to the other project parameters in the form of cost, environment, productivity, quality and schedule (time).

6.1   Tower crane bases, supports, anchors and ties

Tower crane incidents occur globally, and South Africa is no exception. Furthermore, tower cranes are a major item of plant, entailing erection, use and dismantling. For example, a construction worker was injured after a tower crane collapsed on a construction site in Umdloti in Kwazulu-Natal on 14 August 2017. Figure 16.2 shows the collapsed crane.

Figure 16.2: Collapsed tower crane, Umdloti, Kwazulu-Natal (Reaction Unit SA in Mngoma & Pillay, 2017)

Regulation 22, 'Cranes', of the Construction Regulations, 2014 addresses tower cranes, including reference to the Driven Machinery Regulations, 1988 [Updated 24 June 2015].

Key H&S requirements include:

- scientific design
- risk assessment and a resultant method statement prior to initiating the design and erection processes
- consideration of likely wind forces
- scientifically designed, firm, level and secured bases for fixed tower cranes and tracks for rail-mounted tower cranes, including prior geotechnical surveys
- attachment to the structure by crane ties in accordance with manufacturer's instructions
- safe ascent and descent for the operator
- supervision of erection and dismantling by a competent person, appointed in writing
- a fall protection plan
- provision of fall prevention and fall arrest equipment
- evaluation of the riggers' and operator's medical fitness
- erection and dismantling by competent riggers
- operational wind limit warning device
- load limiting device
- inspection after erection by a competent person
- operator pre-use checks before each shift
- operator weekly in-use inspections
- maintenance by competent riggers in accordance with the manufacturer's instructions
- records of all inspections and maintenance
- competent and medically fit operator;
- competent banks people
- competent supervision of all lifting operations.

Considerations:

- a possible need for mobility
- proximity to other tower cranes, including jib heights
- power supply
- maximum loads to be lifted relative to various reaches
- sequencing of construction of the enclosing fabric because of the mast stays

- the structure to which the mast is to be attached
- ergonomics of the cab
- operator/banks people communication
- temporary works manager/operator communication.

6.2 Personnel and material hoists

The Construction Regulations, 2014 address material hoists in Regulation 19, however, they are silent with respect to personnel hoists.

Key requirements include:

- an adequate structure to which the hoist will be attached
- supervision of the design and erection by competent persons, appointed in writing
- a fall protection plan
- evaluation of the riggers' and operator's medical fitness
- provision of fall prevention and fall arrest equipment
- competent riggers
- competent operator
- scientifically designed, firm and level foundation
- adequate intermittent securing of the mast to the permanent structure in accordance with manufacturer's guidelines
- a substantial enclosure at ground level around the hoist way
- suitable gates at ground level enabling access to the platform
- numbered gates providing access to the platform at every level
- an overrun device above the highest platform
- indication of the safe working load on the platform or the permissible number of persons based upon an average weight per person
- inspections and records
- maintenance and records.

Considerations include:

- power supply
- the prevailing wind direction
- the interaction between prevailing wind and the permanent structure and hoist
- number of people to be conveyed per period and at peak times
- nature and mass of materials to be conveyed.

Figure 16.3 provides visual reinforcement of the key requirements and considerations.

Figure 16.3: Personnel hoist, Manchester, United Kingdom (Smallwood, 2019)

6.3 Temporary electrical installations

Regulation 24, 'Electrical installations and machinery on construction sites', of the Construction Regulations, 2014 addresses temporary electrical installations to a limited extent.

Key requirements include:

- protection of workers against contact with electrical cables and apparatus
- durability, to the extent that electrical installations and machinery can withstand construction site conditions
- designation of a competent person to control such electrical installations
- weekly inspection by the competent person, the maintenance of a register and daily inspection of electrical machinery by the authorised operator or user and the maintenance of a register.

However, SANS 10142-1, 'Wiring of premises', Part 1 'Low-voltage installations', section 7.4, 'Construction and demolition site installations' addresses temporary electrical installations on construction sites in detail. The six sections address:

1. general
2. supply

3. protection
4. installation
5. luminaires
6. distribution boards (DBs).

Supply refers to identifying equipment with the supply that energises it. Protection refers to earth leakage protection devices for socket outlets and lighting circuits. Installation refers to:

- the degree of protection to assemblies, and to fixed and other equipment
- no strain to terminations of conductors
- protection of cables traversing roads against mechanical damage
- use of armoured cables or protection of cables where there is a risk of mechanical damage
- no mixed loading of circuits
- effective means of isolation of all poles to electric motors.

Luminaires refers to fixing and guarding luminaires when luminaires are fixed below 2.5 m above floor level. DBs refer to:

- their compliance with SANS 61439-4
- additional earth leakage protection to cables supplying sub-DBs at their origin
- number of sub-DBs and circuits to be determined by the size of the site and the demand for power
- appropriate circuit breakers
- protection of metalwork against corrosion and damage
- orange colour in terms of final finish
- isolating devices for securing in the off position
- LED indicator lamps per phase to show that the board is alive
- requirements for four-pin (three-phase and earth) and three-pin welding socket outlets
- sufficient number of the different types of socket outlets
- cabinet size.

Considerations include:

- the need for single- and/or three-phase power supply
- major plant and equipment
- emergency generator capacity
- power intensive activities
- number of subcontractors
- the demand per floor on a multistorey project, ie number of floors per sub-DB.

6.4 Cantilever loading platforms

Cantilever loading platforms are used around the world in the construction of multistorey buildings to convey materials and small plant and equipment to upper levels, and to remove waste from these levels, as shown in Figure 16.4. They are either fixed platforms, and thus not retractable, or rolling platforms, which means they can be extended and retracted. Many propriety platforms are available globally.

Key requirements include:

- scientific design
- supervision of the design and installation by competent persons, appointed in writing
- integral non-detachable edge protection
- fabrication in accordance with design, if not a propriety platform
- an adequate structure to accommodate the platform
- a fall protection plan
- evaluation of the riggers' medical fitness
- provision of fall prevention and fall arrest equipment
- positioning and securing of vertical restraints
- indication of the safe working load on the outer face of the platform and on the entrance to the platform
- inspections and records
- maintenance and records.

Considerations include: the prevailing wind direction; the position of the tower crane or more; the likely loads to be accommodated and the frequency of use.

Figure 16.4: Cantilever loading platform, Manchester, United Kingdom (Smallwood, 2019)

6.5 Rubble chutes

The removal of waste from double- and multistorey buildings is challenging. The use of cranage to convey a filled skip from, say, a cantilever loading platform is an option, however, it consumes hook time. The use of wheelbarrows in conjunction with a materials hoist is another option. However, a rubble chute is ideal because the waste is then concentrated at a collection point. Construction Regulation 14, 'Demolition work', (6) and (7) prescribes requirements for rubble chutes.

Key requirements include:

- scientific design
- design, supervision of and erection by competent persons, appointed in writing
- degree of closure versus inclination
- a level, firm and stable foundation for the collection skip
- a robust collection skip
- enclosure of the collection skip
- secured linkage of the chute sections
- secure attachment of the chute to the structure
- signage
- a fall protection plan
- evaluation of the riggers' medical fitness
- provision of fall prevention and fall arrest equipment.

Considerations include the fact that the waste cannot be sorted, which constitutes a challenge in terms of environmental management. However, education and training of the workforce, and supervision can mitigate the consolidation of waste types.

7. Bulk Mixing Plants

Bulk mixing plants or batching plants, as they are more commonly referred to, are addressed by Regulation 20 of the Construction Regulations, 2014. Bulk mixing plants have been included as a regulation because they have featured in accidents. The Weekend Post reported on an accident in Parow in Western Cape on 11 March 2011 in which a worker became trapped inside a concrete mixer after sustaining fatal crushing injuries. The report stated that a lack of communication of sorts resulted in the concrete mixer being turned on while the worker was still inside cleaning the concrete mixer. The report stated that this was the second fatality resulting from such an accident in a month in the Western Cape.

Key requirements include:

- supervision of the plant by a competent person appointed in writing who is aware of the dangers and the required precautionary H&S measures
- operation by a competent operator

- erection and placement of the plant in accordance with the manufacturer's requirements
- erection as per design
- start and stop devices must be in an easily accessible position and constructed in such a manner as to prevent accidental starting
- ensuring that all dangerous moving parts of a mixer are beyond the reach of persons by means of doors, covers or other similar means
- only supervisor authorised removal or modification of any guard or safety equipment
- compliance with precautionary measures stipulated for confined spaces in the General Safety Regulations, [Updated to 18 July 2003] when entering any silo, and maintaining a record of all repairs or maintenance.

Considerations include: delivery of materials may require temporary roads, and reach of tower cranes or mobile cranes.

8. Groundworks for Plant Operation

Platforms for cranes and piling rigs are required to provide a reliable and stable surface on which cranes, such as crawler cranes, piling rigs and other ground treatment machinery can operate safely.

Key requirements include:
- scientific design
- substrata investigation
- maximum ground pressure loading
- appropriate platform material and method of placement
- maximum aggregate size
- adequate drainage
- verification of loading capacity
- supervision by a competent person, appointed in writing
- daily inspection and records.

Considerations include:
- characteristics of substrata
- buried services
- backfilled trenches
- soft spots
- pile holes
- presence of groundwater
- access routes to the platform
- grade limitations of the plant

- minimum edge distances to side slopes
- slope stability hazards
- areas for assembly of plant
- operational restrictions, such as slew restrictions or lifting positions, for critical lifts
- pile holes during construction
- possible contamination during use.

9. Review Questions

- Are the Construction Regulations adequate in terms of addressing the range of ancillary temporary works undertaken in construction?
- Is the current level of construction ancillary temporary works education, training and continuing professional development in South Africa optimum?
- Does the Occupational Health and Safety Inspectorate of the Department of Employment and Labour interrogate ancillary temporary works design processes and execution to the necessary extent?

10. Review Exercises

- Compile a 750-word article titled *Assuring healthy and safe ancillary temporary works in general.*
- Compile a flow progress diagram illustrating the common interventions required to assure healthy and safe ancillary temporary works in general.

Bibliography

Mngoma, N & Pillay, K. 2017. One injured in Umdloti crane collapse. *IOL*, 14 August 2017, https://www.iol.co.za/mercury/news/watch-one-injured-in-umdloti-crane-collapse-10785129.

Weekend Post. 2011. Man dies while cleaning mixer. *Weekend Post*, 12 March 2011: 4.

CHAPTER 17

Construction Health and Safety Induction

Theo C Haupt

1. Introduction

The construction industry experiences a disproportionately high number of accidents, mainly because it is labour intensive, employs a large unskilled workforce, is characterised by hazardous work activities, is affected by the demands of heavy duty plant and equipment and is impacted by the constantly changing working environment of a typical construction site. The contribution of induction to the success of a construction health and safety (H&S) management system and engendering a culture of construction H&S in the workers on a construction site is unquestionable. Induction is the practice of providing work-related health and safety information to all parties exposed to construction works, such as new entrants and visitors to the site, or existing construction workers on a construction, before engaging in actual work, to help them settle and become familiar with the work environment. Induction involves familiarising construction site workers with the site-specific H&S hazards that are likely to be encountered while working on the site, so that they may conduct themselves in such a manner that they do not endanger themselves or anyone else. The best laid construction H&S plans cannot be effective unless they are communicated to the workers who will actually be exposed to the hazards on the construction site.

Prevention of accidents by offering induction is one of the most effective means of dealing with hazards on construction sites before they occur. Induction is the starting point for an organisation to introduce a culture norm that supports H&S. Studies suggest that induction on its own is not sufficient to fully discharge all legal obligations in relation to H&S. An effective system should include aspects of planning to tailor site-specific H&S practices, constant supervision, fair and appropriate discipline and continued knowledge and awareness of site risks that construction workers are exposed to as the site changes. Furthermore, experience suggests that high H&S standards are achieved on projects where the construction clients are committed to H&S themselves and provide appropriate management oversight.

Construction workers also have a role to play in ensuring that they implement what is learnt and disseminated during induction. Construction H&S induction is important because all construction sites are different and have a wide range of hazards that will change as the construction project progresses. Construction H&S induction must, therefore, be specific to a particular site and provide information on the current hazards pertaining to the construction site and the applicable site rules. Induction is not intended to take the place of construction site-, activity- or task-specific instruction, training or supervision.

2. Achieving Effective Construction H&S Induction

Because of their relationship with their work crews and their influence on them, construction supervisors must have the competence to be able to conduct an H&S induction that communicates the organisation's expectations of H&S to workers on construction sites. To achieve effective site H&S induction and a positive H&S culture, supervisors must be able to use modern technologies and multilingual skills, given the multicultural and multilingual profile of construction workers in South Africa. Induction should be done in a variety of languages and induction material should be translated into a variety of languages, and should developed with a sensitivity towards cultural needs.

Inadequate construction H&S induction can have serious consequences that include threats to profitability, corporate social responsibility and increased turnover of the construction workforce, leading to poor productivity. Negligent and non-adherent attitudes to construction H&S induction will make preventive measures for the reduction of accidents on construction sites difficult to implement. Adequate induction can improve the health and well-being of construction workers, reduce financial implications on construction firms and encourage productivity on construction sites. Proper induction practices allow the construction project to be completed within its targeted execution time, without delays caused by lost labour hours by construction workers who have died or been injured as a result of site accidents.

Construction H&S induction should be planned to take into account an overview of the general legal requirements, to provide an overview of H&S requirements for the construction site, to inform construction employers and construction workers of their legal rights and obligations, to help project stakeholders to identify hazards on the construction site, to assess risks and to implement controls to reduce the likelihood of construction workers being injured.

3. Benefits of Construction H&S Induction

Construction H&S induction is a necessary and important intervention that has several positive influences and benefits, which include the following:

- demonstrates the involvement and commitment of management to construction H&S
- construction workers understand their H&S obligations and responsibilities on the project and what sanctions are given if they violate the construction project H&S rules
- communicating the results of risk analyses that were undertaken to identify H&S risks on the construction site
- garner a greater awareness of construction H&S obligations of all parties, especially a duty of care, hazard assessment and identification and manual handling
- improvement of the construction H&S culture at all levels of construction activity on the site
- provide an understanding of the importance of construction H&S to the project
- provides the latest information about conditions on the project site because the conditions can change daily
- communicates information about the applicable construction H&S regulations
- provides information about construction work procedures in the actual construction work environment.

4. Types of Construction H&S Induction

There are three types of construction H&S induction, namely general induction, construction site induction and construction activity-specific induction.

4.1 General induction

General H&S induction provides basic knowledge and principles of H&S to new workers about the common hazards that could be encountered and how they could be addressed. It provides a broad general appreciation of H&S induction aspects without detailed specific technical aspects, especially considering that the majority of the labour force is unskilled and may not relate to the detailed technical specifics.

4.2 Construction site induction

Site induction is conducted on-site by individual contractors and, therefore, focuses on the rules and regulations that are support the contractor's method statement. It addresses the H&S issues and safe work practices that are specific to a particular building and construction site, which include knowledge of the contractor's rules and procedures for site H&S, emergency management, supervisory and reporting arrangements and other site-specific issues.

4.3 Construction activity-specific induction

The aim of construction activity-specific induction is to provide construction workers with knowledge of the H&S issues and safe work practices relating to a specific construction activity or trade on site. As a construction project progresses, there is a need to inform affected workers about the potential hazards that they may be exposed to. This type of induction is done specifically for special operations that require operational appraisal or activities that are high risk, such as working at heights. For example, when new plant and equipment are brought to the construction site, specific induction needs to be carried out to inform the workers who will use this machinery of the hazards that they may be exposed to. For construction activity-specific induction the following must be addressed:

- identifying the construction activity that is regarded as a high-risk construction activity
- specifying hazards relating to the identified high risk construction activity and the risks to the health and safety of everyone involved in the project, especially construction workers
- describing the interventions and measures to be implemented
- describing how these control measures will be implemented, monitored and reviewed on the construction site.

5. Content of Construction H&S Induction

The content of the induction program should include, and stress, the construction activity-specific and current hazards, rather than generic hazards that the construction workers often feel familiar with, such as construction site rules, site-specific hazards, risks to workers performing their jobs and procedures for reporting incidences and injuries. Induction should, therefore, include information on the following aspects:

- management commitment to and involvement in construction H&S
- information about the construction site, specifically nature of the work and site-specific hazards
- specific hazard exposure management
- explanation of the reasons for and philosophy behind the construction H&S policy and each H&S rule
- the duty of care
- H&S structures, such as H&S representatives and committees
- nomination and introduction of first-aider(s) and the location of first aid stations
- work methods and construction activities

- presence and proper use of plant and equipment
- safety signage and its meaning
- available welfare facilities to be used
- site emergency and evacuation procedures to be followed
- environmental issues
- housekeeping requirements
- waste management and disposal
- proper use and care of personal protective equipment (PPE) as a measure of last resort after efforts have been made to mitigate the frequency and severity of any likely exposure to risk
- arrangements for reporting of accidents and near misses
- details of any responsibilities of employers and workers determined by general and construction H&S legislation and associated regulations
- scope of the authority of workers determined by job descriptions, including details of when they have the right to stop work activities
- H&S inspections
- details of any planned training and toolbox talks
- providing awareness of relevant regulations relating to construction H&S on the site
- the right of workers to refuse to undertake hazardous construction work activities.

Including these aspects of construction induction will help to reduce the occurrence of accidents on construction sites and protect construction workers and visitors from the hazards associated with construction activities, while disseminating important information on how to prevent accidents and adjust to a changing work environment. Example 17.1 provides some details of what should be included in an effective construction H&S induction programme, as advocated by the Health and Safety Executive in the United Kingdom.

Example 17.1: Some inclusions in a construction H&S induction

Element	Example	Comment
Welcome and introduction		
Commitment of the company to H&S	We are committed to protecting the health and safety of all people working at or visiting our construction site. We plan, manage, conduct and supervise all our work in compliance with legislation and better practice. We want to ensure that all construction workers have a clear understanding of their responsibilities, along with the responsibilities of the company.	Describe how the company H&S policy works and the roles people play in terms of the policy. Stress the legal duty of everyone to contribute to healthy and safe working and a healthy and safe workplace.

$\rightarrow$

Element	Example	Comment
Importance of construction site induction	You have probably gone through many construction site inductions and will probably go through many more. The induction is important as all construction sites are different and present a wide range of hazards, which will change as the construction project progresses over time. This construction site induction is specific to this site and provides you with information on the current hazards on the site and tells you about the site rules. Please pay attention for the next few minutes.	Explain to everyone the importance of construction site inductions as all construction projects and sites are different and present a wide range of hazards, which will change as the construction project progresses over time. Explain that this construction site induction is specific to the particular construction site. Ask everyone to pay attention for the next few minutes.
Project-specific conditions/ requirements	The project: ■ project history ■ current stage ■ future programme of work ■ type of construction ■ end use and clients requirements ■ location of statutory notices.	Explain to everyone the requirement to observe site-specific elements appropriate to their own work activities and/or site-wide hazards. These may include but not be limited to the following: ■ open excavations ■ work at height ■ overhead power lines ■ confined spaces ■ contaminated land ■ excessive vehicle movement ■ traffic management systems ■ fire risks. Ensure that everyone is made aware of specific requirements for the production of risk assessments and method statements where specific hazards are identified. Make everyone aware of areas of work that will require specific authorisation to proceed, such as a permit to work. Ensure everyone is made aware of restricted areas and the reasons for the control measures in place.

$\rightarrow$

Element	Example	Comment

Emergency evacuation and fire	The following must be covered during the induction:
	■ the alarm
	■ exit routes
	■ assembly points
	■ fire points
	■ fire prevention, such as:
	  — no smoking and be aware of other possible ignition sources
	  — keeping the site tidy , meaning less material to burn
	  — hot work permits.
		Describe the type of situations that might require evacuation of the site. Have a site layout plan available.
		
		Ensure that everyone knows what the on-site alarm sounds like and how to raise it.
		
		Ensure that everyone knows the different routes (KEEP THEM CLEAR) that they may have to use to leave the site and where they should assemble for roll call.
		
		Ensure that everyone knows where firefighting equipment is situated, that they are trained to use any firefighting equipment, that they should only attempt to fight small fires, that they should have a clear escape route and use only after the alarm has been raised.
		
		Stress prevention and give examples of ignition sources.

Details of construction site staff	You must be able to identify the: ■ site contracts manager ■ engineers ■ supervisors ■ safety representative ■ safety officer ■ first-aiders ■ fire marshal ■ the site contact telephone number.	Ensure that workers know who to contact and how to reach them. Amend list as required. Ask the workers to write down the information on this list or provide a printed sheet.
Who is on site?	Please ensure you sign in and out of the site; if you do not, we will not know you are here for security and emergency purposes.	Ensure that all inductees are made fully aware of the site procedure for recording who is on site at any given time. Explain to them that the main purpose of this is to ensure that all persons are accounted for in the event of an emergency situation, not as a timekeeping tool. Explain that any person not accounted for in an emergency will be treated as missing, which may put emergency workers at risk if they have to look for someone who is not actually on site. Explain that failure to comply with this requirement may result in disciplinary action being taken against offenders.

$\rightarrow$

Element	Example	Comment
Welfare facilities	Find out where these facilities are located: ■ toilets ■ canteens ■ first aid.	Show where facilities are on site plan. Reinforce the fact that, while management has the responsibility to provide suitable and sufficient welfare facilities, workers have a responsibility to look after these facilities and report any damage or vandalism. Leave toilets and canteens as you would wish to find them. Be aware of first aid procedures for the site and who to contact or where to go for help.
Housekeeping	Always clear up your own rubbish: 'A clean site is a safe site.'	Stress the importance of good housekeeping to prevent slips, trips and falls, and remove material that could fuel fires. A high number of accidents occur on construction sites because workers trip over rubbish and waste. Failing to ensure a tidy workplace is unsafe and could cause the site to be closed down. Cover any waste management requirements, for example, location of skips and disposal requirements. Discuss any environmental issues pertaining to the site, for example, chemicals, smoke, noise or even wildlife.
Environment and waste disposal	All waste should be disposed of in the correct skips: ■ Under no circumstances should liquid waste, such as paints or solvents, be allowed to soak into the ground or be poured down drains. This is 'hazardous waste' and should be disposed of in line with current legislation. ■ Bonfires must not be conducted on site.	All waste should be disposed of in the correct skips. Under no circumstances should liquid waste, such as paints or solvents, be allowed to soak into the ground or be poured down drains. This is 'hazardous waste' and should be disposed of in line with current legislation. Bonfires must not be conducted on site.
Drugs and alcohol	■ Any persons caught in possession or under the influence of drugs or alcohol will be removed from site. ■ If you are on drugs for any medical reason, please inform your supervisor at once.	Explain to everyone the importance of reporting for work in a fit state. It is not only their safety that is at stake but those around them who could be affected by what they do. Explain that any person reporting for duty under the influence of drugs or alcohol will be treated very harshly and that removal from site is not just a threat.

→

Element	Example	Comment
Drugs and alcohol (cont.)		Make everyone aware of the substance abuse policy (if appropriate) and inform them of the procedures for testing and counselling. Ask everyone whether any of them is on medication for a specific medical condition that may impair their performance to make this known to their manager so that appropriate measures can be put in place.
Working at height	You must: ■ use secure platforms with proper edge protection ■ protect holes, leading edges and fragile materials ■ consider weather conditions. If in doubt, speak to your supervisor.	Stress working at height is the biggest single cause of construction fatalities annually, making up 40–50% of all construction fatalities annually. Describe the kind of height work on the site and the controls required using appropriate slides to stress and illustrate the main points.

Example 17.2 provides an example of a construction H&S induction record, which should be kept on the construction site.

Example 17.2: Construction H&S site induction record

EXAMPLE OF CONSTRUCTION SITE INDUCTION RECORD

Project Name:	
Project No.:	

PERSONAL DETAILS:

Surname:		First Name:		DOB:	
Address:					
Home Number:		Mobile Number:			

NEXT OF KIN:

Next of Kin:		Relationship:	
Contact Number:			

EMPLOYMENT:

Employer:		Trade:	
Supervisor:			

$\rightarrow$

HEALTH:

Any medication, allergies or illness – **give details**:	

TRAINING/INDUCTION REQUIREMENTS/LICENCES/COMPETENCIES (*attach copies*)

Tick (✓) applicable							
First aid		Paving breaker		Plant operator		Other	
Demolition		Confined space		Scaffolder			
Jackhammer		Scaffolding		Explosives			
Crane		Working at heights		Driver's license			

INDUCTION REQUIREMENTS Tick (✓) where applicable, N/A for items Not Applicable			
Advise of construction health and safety management plans, policies and emergency response procedures		Hygiene and housekeeping	
Duty of care – employer and employee		Reporting hazards, accidents and injuries	
General site H&S rules and site-specific hazards		Construction site permit requirements	
Show emergency contact numbers/first aid		Excavation	
Mandatory PPE and clothing to be worn on site		Hot work	
Scaffolding/platform ladders		Concrete/floor cutting and core drilling	
Evacuation/muster point/first aid box		Confined space entry	
Availability of legislation (where and how accessed)		Plant inspections and plant risk assessments	
Site vehicles		Cranes and mobile plant	
Discuss your work tasks on site		Material safety data sheets (MSDS) and hazardous substance risk assessments	
Safe working procedures (SWPs) required, signed off by all and must be site-specific		Electrical safety and tagging	
Check and copy licenses required for work		External assistance provider (EAP)	
Communications – toolboxes, pre-project etc		Start and finish times	

INDUCTION Points to Raise

1. What is the minimum personal protective equipment and clothing requirements on this site?
2. Where is the nearest first aid box located?
3. To whom do you report safety and health issues or hazards?
4. What must you do if you need to alter any scaffold?

→

WORKER ACKNOWLEDGMENT

I acknowledge that I have received and will read the XXXX induction booklet. I will ask my supervisor or the site manager if there is anything in the booklet that I do not fully understand.

I will also:

- Comply will all XXXX requirements, processes and procedures.
- Wear the required personal protective equipment relevant to my workplace.
- Report any safety, health and environmental hazards I become aware of.
- Consult and cooperate with XXXX management on safety, health and environmental matters.

Construction worker:

Name	Title	Date

Induction presenter:

Name	Title	Date

The following is an example of a construction H&S induction.

Example 17.3: Example of a construction H&S induction

XYZ CONTRACTORS

CONSTRUCTION HEALTH AND SAFETY INDUCTION

A. General Requirements

All contractors shall ensure that their workers receive construction health and safety orientation prior to starting work on this project.

Each contractor shall maintain, and make available for inspection, records of such construction health and safety orientation and training.

The orientation shall follow the written format specified on the attachment on the next page, in addition to any job-specific information.

All contractors shall ensure that each employee receives a copy of this orientation/induction and signs the acknowledgement page at the end.

B. Construction health and safety orientation or induction

It is our intention to provide and maintain a totally healthy and safe construction site. Your commitment to construction health and safety is a condition for continuous employment on this construction project.

After you have reviewed these guidelines, sign the last page where indicated and return that page to your superintendent or foreman.

$\rightarrow$

1. **Evacuation**: In the event of a fire or any time where project evacuation is required, all workers onsite will be informed via a radio signal or other method as designated by the client or the designated representative of the client.

    YOU SHALL IMMEDIATELY:
    - Stop all work and shut off all electrical equipment, including welding machines and air compressors etc
    - Close valves on gas cylinders
    - Walk! (DO NOT RUN OR JUMP FROM ELEVATED POSITIONS) to the designated assembly points. Remain at the assembly point until the all-clear signal is sounded. Be prepared to follow the directions from your supervisor.

2. **First aid**: All injuries are to be reported to the principal contractor's representative immediately. DO NOT LEAVE THE SITE WITHOUT REPORTING AN INJURY, REGARDLESS HOW MINOR YOU MAY THINK IT IS.
    - Injuries requiring attention and care of a medical doctor will require a drug screen and a medical authorisation form from your supervisor.
    - If we have a construction worker injured on our construction site, we want the best medical care possible. However, if we have an injury that we suspect is fraudulent, we will spare no expense investigating and prosecuting.

3. **Personal Protective Equipment**:

    Head Protection: A hardhat must be worn at all times upon entering the construction work area or site, with the bill to the front.

    Eye and face protection: Appropriate eye protection (American National Standards Institute (ANSI) Z87)), with side shields, is required to be worn by all workers on the construction site at all times in areas where indicated by appropriate signs and the construction activity requires it. Prescription glasses must be approved safety glasses, approved glasses and frames or approved eye protection.
    - When grinding or buffing, a face shield with approved safety glasses will be required.
    - When cutting or burning, safety goggles will be required.
    - When welding, a welding hood and lens with an appropriate number filter must be worn.
    - Chemical goggles are required to be worn when working with any corrosive or toxic material.

    Respiratory and hearing protection: Respiratory and/or hearing protection is required in designated areas and/or when performing specific construction activities.
    - Construction workers must be clean-shaven before to using a respirator.

4. **Barricades**:
    - Barricade tape is not to be used in lieu of physical barricades for floor, hole or wall openings, or when permanent handrails have been removed.
    - Yellow barricade tape indicates caution must be taken when approaching or entering the construction area.
    - Special authorisation to enter a construction area cordoned off with red barricade tape is required. Anyone entering area without authorisation will be subject to disciplinary action.

5. **Fall protection/tie-off**:
    - A 100% tie-off policy is in effect whenever you are exposed to the potential of falling more than 1.8 metres to a lower level.
    - An approved fall arrest system will be worn when working at unprotected elevations greater than 1.8 metres and when working in powered person-lifts.
    - ·An approved fall-arrest system consists of a full body harness and two shock absorbing lanyards, each with double-action or positive locking snap hooks.
    - Any lifeline, safety harness or lanyard having been subjected to fall loading shall be removed from service.

$\rightarrow$

6. **Lockout/tagout**: Lockout/tagout the power source prior to making adjustments or repairs to any equipment. DO NOT DEPEND on the control switch on drills, grinders etc. UNPLUG THEM.

7. **Electrical tools and cords**:
   - Tools are to be visually inspected by the construction worker prior to use. Any tool or cord found to be defective should immediately be taken out of service.
   - Use approved ground fault circuit interrupters for all temporary wiring that is not part of the permanent wiring of the building or structure.
   - When working in an existing building where power is not protected by ground fault circuit interrupters, the contractor shall supply and utilise in-line (pigtail) ground fault circuit interrupters.
   - Use an assured grounding conductor programme in tandem with all ground fault circuit interrupters.
   - Check the RPM rating of grinding wheels or discs and ensure that the RPM rating is greater than that of the driver.
   - Do not alter or adapt tools and guards.
   - Maintain electrical cords and welding leads at a 2.1 metre length, avoiding pinch points and creating trip hazards.
   - Do not tie electric cords to metal rods or nails.

8. **Ladders**:
   - Ladders must be free from defects.
   - Place the ladder so that its base is out ¼ (one-quarter) the distance of the height to be worked at.
   - Tie ladders at the top or secure at the base.
   - Do not extend an extension ladder to its full length; overlap at least 3 rungs.
   - Do not use stepladders as extension ladders.
   - Fully extend stepladders and lock them in position.
   - Only one worker at a time shall work off a stepladder.
   - Do not stand or sit on the top or top two rungs of a stepladder.

9. **Scaffolds**:
   - Completely deck all scaffolds, platforms, and staging, with decking, secured and built with standard handrails and toe boards on the open sides and ends.
   - The footing for scaffolds shall be sound and capable of carrying the maximum intended load.
   - Do not erect, move, dismantle or alter any scaffold, except under the supervision of a competent person.

10. **Explosive actuated tools**: Employees must be trained/certified before they may use these tools.

11. **Clothing**:
    - Construction workers will work fully clothed.
    - All employees shall wear proper safety footwear while on the project. Some construction activities may require additional foot protection.

12. **Jewellery**: Use good judgment as to the type of jewellery that will not constitute hazard. For instance, earrings or chains that could get caught in construction tools and machinery are not allowed.

13. **Compressed gas cylinders**:
    - Cap, tie-off or otherwise properly store compressed gas cylinders when not in use.
    - Cylinders must remain in an upright position at all times.
    - Keep protective caps in place at all times.
    - Do not use oil or grease on valves or gauges.
    - When in storage, separate oxygen cylinders from fuel gas cylinders by at least 7 metres, or by a 1.5 metre wall with a half-hour fire rating.

→

14. **Lifting**: Like everything else, the right way to lift is easier and safer. If the load is too heavy, GET HELP. Do not lift with your back, bend your knees.

15. **Lifting and/or swinging loads**:
    - Do not walk under a suspended load or permit other workers to do so.
    - Barricade the lift area to control access into the area.
    - Never pick up a load in excess of the capacity of the equipment.
    - Only one worker at a time will give hand signals to crane or lift truck operator.
    - Use tag lines to control loads.
    - Never leave a suspended load unattended.
    - Never ride on a load, crane hook, headache ball, or forks of a lift truck.

16. **Rigging**:
    - Never use hands or feet to guide cable or line onto a drum or hoist. Use a bar as a guide.
    - When it is necessary to stretch cables or lines across roads or walks, block the road or walkway if the cable or line is lower than 5 metres above roads or less than 2.1 metres above walkways.
    - Seat chain links into a hook by hand pressure only. Never hammer a chain link onto a hook.
    - Use the approved method to fasten hoisting equipment together.
    - Follow the manufacturer's recommendations to determine the safe working loads of hooks. Test all hooks for which no applicable manufacturer's recommendations are available to twice the intended safe working load before they are initially put into use.

17. **Chain blocks**: When using chain blocks, inspect and check for proper operation using a test load before making a critical lift.
    - Know how much you are lifting and the chain block limitations.
    - Only one person at a time shall pull on the chain of a block.
    - Never use a load chain as a sling for lifting.
    - Do not use chains for rigging purposes, with the exception of chain falls with the capacity plate intact.
    - Straighten chains and make every link seat before lifting. Never jerk or put any strain on a kinked chain.
    - Use appropriate or rated material to suspend or anchor chain blocks.

18. **Equipment operations**:
    - All equipment operators must be trained for the type of equipment being operated. The contractor shall provide proof of competency for all individuals operating heavy equipment.
    - No passengers are allowed to ride on equipment with operators.

19. **Access**: Climbing and/or sliding down columns or diagonal bracing are not permitted. Walking on elevated beams and pipes without being tied off is not permitted.

20. **Permits**: There are various permits required on the project. The principal contractor shall issue the appropriate permits and maintain records. Commonly used permits include:
    - Hot work: Any work, tool or equipment, such as welding, burning, grinding, vehicles and portable welders, that might provide a source of ignition in areas where combustibles are present.
    - Confined space: The authorisation required to enter any vessel, pipe, confined space or excavation for any reason.
    - Lock and tag: Prevents the operation of a valve, switch or piece of equipment when injury or property damage could result from the operation.

$\rightarrow$

- Scaffold: Permission to use a scaffold that has been erected. A scaffold permit shall be secured by each new contractor or subcontractor who needs to use a scaffold, following a review of their proposed operation.

Failure to follow the instructions on a tag or permit will constitute grounds for removing the construction worker from the construction site. If you see a tag that you do not understand, ask your supervisor.

21. **Hazard communication**: Handling and storage of chemicals are the two most common causes of accidents. Information will be provided to every construction worker in the following ways:
    - Container labelling: Labels give you information about the immediate hazards associated with the chemical.
    - Material safety data sheets (MSDSs): MSDSs give you detailed information about the chemical, such as the associated physical and health hazards, first aid, firefighting, and personal protective equipment required.

Know what you are handling, read the label and, if there is any doubt, consult the MSDS.

22. **Parking and motor vehicles**:
    - All workers must park their personal vehicles in the designated areas only.
    - Posted regulations governing the use of the parking areas must be strictly followed.
    - All vehicles parked on the construction site will be at the risk of the vehicle owner and the company accepts no responsibility for damage to, or theft of or from such vehicles.

23. **General**:
    - Only drink water from approved drinking water containers or dispensers.
    - Proper housekeeping is essential and will be part of every job.
    - Clean up all spills or leaks promptly. The contractor is responsible for containing and cleaning up all spills caused by its construction workforce.
    - Obey all posted speed limit signs on the construction site.
    - Pedestrians will always have the right of way.
    - Always yield the right of way to emergency vehicles.
    - Smoking is permitted in designated areas only and not in the construction work area.
    - No firearms or weapons are allowed on the construction site.
    - Riding on any equipment that is not designed for worker transport is prohibited.
    - Ride in vehicles with seats firmly attached.
    - Construction workers must obey all danger and caution signs.
    - When possible, correct all unsafe and unhealthy conditions. Report all unsafe conditions to your immediate supervisor or health and safety representative.
    - No running is permitted on the construction site.
    - All material raised and lowered from any height must be done by rope. No dropping or throwing of material from height is permitted.
    - No horseplay will be tolerated.
    - No fighting. All workers who are involved in a fight will be removed from the construction site and may be reported to the South African Police Service.

6. Review Questions

- Why should construction workers attend an induction session?
- What are the benefits of construction H&S site induction?
- What should be included in a construction H&S induction programme?
- Describe each of the types of induction.

7. Review Exercise

- Draft a short induction programme for the erection of scaffolding, material and personnel hoist and housekeeping for the project shown in the following photograph.

Bibliography

Bahn, S & Barratt-Pugh, L. 2014. Safety training evaluation: The case of construction induction training and the impact on work-related injuries in the Western Australian construction sector. *International Journal of Training Research*, 12 (2): 148–157.

Bahn, ST. 2012. Construction induction training: Does mandatory training work? *Journal of Health, Safety and Environment*, 28 (3): 17.

Fauzi, AZ, Siswanto, AB & Salim, MA. 2020. Effect of safety induction, reward, and punishment on K3 discipline (Case study: USM Tower Project). *International Journal of Advanced Research in Engineering & Management*, 6 (4): 2456–2033.

Ganah, AA & John, GA. 2017, BIM and project planning integration for on-site safety induction, *Journal of Engineering, Design and Technology*, 15 (3): 341–354.

Griffith, A & Howarth, T. 2014. *Construction health and safety management*. Harlow: Pearson Education, Inc.

Hashim, SZ, Ahzahar, N, Bayani, I & Hashim, Z. 2018. Safety induction program in relating of enhancing safety awareness at construction site. *Advances in Natural and Applied Sciences*, 12 (8): 16–20.

Health and Safety Commission. 2007. *Managing health and safety in construction: Construction (Design and Management) Regulations 2007: Approved code of practice.* Sudbury: Health and Safety Executive Books.

Hinze, J. 2006. *Construction safety.* Gainesville: Jimmie Hinze.

Holt, ASJ. 2005. *Principles of construction Safety.* Osney Mead, Oxford: Blackwell Science Ltd.

Mutwale-Ziko, J, Lushinga, N & Akakandelwa, I. 2017. An evaluation of the effectiveness of health and safety induction practices in the Zambian construction industry. *International Journal of Health and Medical Engineering*, 11 (3): 614–618.

National Association of Home Builders (US). 2007. *Home builders' safety program.* Washington: Builderbooks.

Zulu, E & Haupt, TC. 2016. An exploratory study of the health and safety induction practices on a major construction site in South Africa. *Proceeding of the 10th Built Environment Conference*, Port Elizabeth, South Africa, 31 July–2 August 2016.

CHAPTER 18

Construction Health and Safety Method Statements

Theo C Haupt

1. Introduction

In construction, method statements are common practice and form part of the construction management and supervision processes on almost all construction projects. According to the Health and Safety Executive in the United Kingdom, a construction health and safety method statement (CHSMS) is referred to as a description, in a logical sequence, of exactly how a job is to be carried out in a safe manner and without risks to health. In this statement, reference is made to all the risks identified in the risk assessment and the measures needed to control those risks, which allows the job to be properly planned and resourced. The CHSMS is a document that details the key activities to be performed in order to reduce, as reasonably as is practicable, the hazards identified in any risk assessment. The method statement will provide additional information to construction workers about how to do the construction activity, what order to follow and the precautions that need to be taken at each step of the activity. Method statements provide a timeframe and sequence. It is, therefore, important to know what the control measures are and also the order in which each construction activity has to be carried out.

Given the poor historic H&S performance record of the construction industry, method statements for high risk construction activities introduce higher levels of planning to ensure that the all the risks are considered before the activity is actually executed on site. While a H&S method statement is not intended to be a procedure, it is a tool to help both construction supervisors and workers confirm and monitor the control measures required for carrying out a construction activity on site.

2. Preparing the CHSMS

The CHSMS must be prepared before the construction activity is carried out on site as it will plan out the construction activities in a logical sequence, providing information on the controls and precautions identified in the risk assessments required at each step. Therefore, before any CHSMS can be prepared, a risk assessment must first be done as it will form the basis of the method statement. On a construction project the party responsible for preparing the CHSMS should be the party executing the construction activity. Therefore, consultation between the principal contractor (PC) and other contractors is necessary to determine who should prepare the CHSMS. Thereafter, construction supervisors, construction H&S representatives and construction workers should be involved to ensure that, when the CHSMS is implemented on site, they understand the detail and what is required to implement it in order to maintain the predetermined risk controls.

The PC must ensure that each subcontractor submits a completed CHSMS to be included in the construction H&S file before the affected work commences.

3. Content of the CHSMS

The CHSMS must:
- identify the construction activity that is regarded as a high risk construction activity
- specify hazards relating to the high risk construction activity and the risks to the health and safety of everyone involved in the project especially construction workers
- describe the interventions and measures to be implemented according to the hierarchy of H&S control to manage the risk exposures as identified during the hazard identification and risk assessment process
- describe how these control measures will be implemented, monitored and reviewed on the construction site.

While the CHSMS must be based on a risk assessment, the risk assessment need not be included in it. However, the risk assessment may be required by a construction H&S auditor or inspector of the Department of Employment and Labour.

4. Features of a Good CHSMS

The CHSMS should be short and concise. It should focus on describing the identified hazards and the control measures that must be put in place to ensure that a high risk construction activity is carried out safely. It must be easy for all workers to understand and may include pictures or diagrams to make communication of the information more effective. CHSMSs should be kept where the construction activity occurs and, if not, then in an easily accessible location on the construction site. While it is not a legal requirement to have a CHSMS for every construction activity on a construction activity, the Construction Regulations, 2014 specifically require a CHSMS when the construction project involves demolitions and the use of explosions during demolitions. Demolition work means work to demolish or dismantle a structure or part of a structure that is load-bearing or otherwise related to the physical integrity of the structure. Having CHSMSs for other activities, apart from these two mandatory ones, is a demonstration of better practice on the part of the PC.

Furthermore, a good CHSMS will follow the principles of prevention to exposures to hazards on construction sites, namely:

- Avoiding the risk altogether
- Evaluating unavoidable risk exposures
- Combating the risks at source, where possible
- Considering the adaptation of the work activity and workplace to the individual construction worker, taking into account the construction method and tools, plant and equipment
- Adapting to and implementing technological alternatives and advancements
- Substituting hazards with non-hazardous or less hazardous options
- Developing and implementing a coherent and comprehensive overall hazard and risk exposure prevention programme on the construction site
- Considering collective protection measures over individual protection measures
- Giving clear and unambiguous instruction to construction workers.

5. Benefits of a CHSMS

Because they may be required to satisfy the tender or bidding requirements of the client, CHSMSs can be useful as estimators to price for major construction elements with high risk. CHSMSs provide the basis for a structured and effective approach to the construction management processes of planning; financial management and control; site establishment and layout; and project scheduling. They provide the basis for improved and transparent communication between all project stakeholders, including construction workers.

6. Format of a CHSMS

While no format is prescribed in the Construction Regulations, 2014, a schedule or pro forma, such as suggested by Safe Work Australia, can be used, as in the example provided in Example 18.1. They should at least include the following:

- a description of the construction activity, including quantity and location
- anticipated construction method
- a detailed sequence of construction activities
- control measures and supervisory arrangements
- resources, such as labour, and plant and equipment, needed to carry out the construction activity safely, including size, weight, power rating and necessary certifications

- time (duration)
- significant hazards and risks
- training requirements
- special considerations including limitations
- details of all personal protective equipment (PPE) and other measures, such as signs and barriers
- environmental limitations.

Example 18.1: Example of the layout of a CHSMS

- The construction activity described must be performed strictly in accordance with this CHSMS.
- This CHSMS must be kept and be available for inspection until the high risk construction work to which this CHSMS refers has been completed. If the CHSMS is revised, all versions should be kept and the revised version should be clearly marked as a revision. The latest version must always be used.
- If a notifiable incident occurs in relation to a high risk construction activity in this CHSMS, the CHSMS must be kept for at least two years from the date of the notifiable incident.

Project name:
Project reference:

Construction project manager: Contact phone:		Date CHSMS provided to PC if by a subcontractor:	
Project description:		Project location/address:	

High risk construction work:	☐ Risk of a person falling more than 2 metres	☐ Work on a telecommunication tower	☐ Demolition of loadbearing structure
	☐ Likely to involve disturbing asbestos	☐ Temporary loadbearing support for structural alterations or repairs	☐ Work in or near a confined space
	☐ Work in or near a shaft or trench deeper than 1.5 metres or a tunnel	☐ Use of explosives	☐ Work on or near pressurised gas mains or piping
	☐ Work on or near chemical, fuel or refrigerant lines	☐ Work on or near energised electrical installations or services	☐ Work in an area that may have a contaminated or flammable atmosphere
	☐ Tilt-up or precast concrete elements	☐ Work on, in or adjacent to a road, railway, shipping lane or other traffic corridor in use by traffic other than pedestrians	☐ Work in an area with movement of powered mobile plant
	☐ Work in areas with artificial extremes of temperature	☐ Work in or near water or other liquid that involves a risk of drowning	☐ Diving work

$\rightarrow$

Person responsible for ensuring compliance with CHSMS :		Date CHSMS received:	
What measures are in place to ensure compliance with the CHSMS ?			
Person responsible for reviewing CHSMS control measures:		Date CHSMS received by reviewer:	
How will the CHSMS control measures be reviewed?			
Review date:		Reviewer's signature:	

What are the tasks involved?	What are the hazards and risks?	What are the control measures?
List the work tasks in a logical order.	Identify the hazards and risks that may cause harm to workers or the public.	Describe what will be done to control the risk. What will you do to make the activity as safe as possible?
Name of Worker(s)		Worker signature(s)
Date SWMS* received by workers:		

* safe work method statement

7. Review Questions

- Who should be responsible for the development of a CHSMS?
- What are some of the benefits of having a CHSMS?
- What needs to be done before a CHSMS can be prepared?
- What should be included in a CHSMS?

8. Review Exercise

- Draft a CHSMS for the erection of scaffolding for the project depicted in the following photograph.

Bibliography

Griffith, A & Howarth, T. 2014. *Construction health and safety management*. Harlow: Pearson Education, Inc.

Health and Safety Commission. 2007. *Managing health and safety in construction: Construction (Design and Management) Regulations 2007: Approved code of practice.* Sudbury: Health and Safety Executive Books.

Holt, ASJ. 2005. *Principles of construction safety*. Osney Mead, Oxford: Blackwell Science Ltd.

National Association of Home Builders. 2007. *Home builders' safety program.* Washington: Builderbooks.

Safe Work Australia. 2014. *Safe work statement high risk construction work.* Information Sheet. Canberra: Safe Work Australia.

Safe Work Australia. 2018. Demolition work – Code of practice, https://www. safeworkaustralia.gov.au/system/files/documents/1705/mcop-demolition-work-v4.pdf.

CHAPTER 19

Construction Health and Safety Safe Working Procedures

Theo C Haupt

1. Introduction

A construction health and safety (H&S) safe working procedure (SWP) is a detailed procedure that breaks down an activity into tasks and simple steps, analyses the potential hazards relative to each step and details how to avoid them. SWPs are directions on how all hazardous construction activities are to be carried out safely on the construction site. They identify hazards and clarify what must be done to eliminate or minimise risks. An SWP is the result of organising tasks in the best sequence of steps to make the best use of people, equipment, tooling and materials, as an important internal control. An SWP or safe operating procedure (SOP), as it is also referred to, is a step-by-step set of instructions that help construction workers perform their tasks in a consistent manner. To put it simply, an SWP documents how a construction activity must be done and forms part of the construction project documentation. It is particularly important to have a written SWP for a complex construction activity that any construction worker can execute with some training, which must conform to regulatory standards, such as erection of scaffolding.

2. A SWP/SOP is a Legal Requirement

A construction H&S SWP is required in terms of the 2014 Construction Regulations in South Africa. Specifically, Construction Regulation 7(1) states that risk assessment requires contractors to include a documented plan of SWPs in the project and site-specific construction H&S plans to be used on construction projects.

3. Compiling SWP/SOP Procedures

The requirements for SWP/SOPs and their format will, in most cases, involved the following steps:

Step 1: Identify the construction activity. Typically the construction activity for which the SWP will be developed will be one that is somewhat complex and hazardous.

Step 2: Define the goal. When you know what the SWP is meant to accomplish, it is easier to write an outline and define the details of the SWP.

Step 3: Involve the stakeholders, especially the construction workers. Converse with anyone who will be engaging in or be impacted by the construction work activity and get them to participate in creating the SWP.

Step 4: Spend some time observing the construction activity, while noting the sequence of the various tasks involved and what needs to be improved to ensure that each step in the sequence can be done correctly and safely. Describe who is involved and what they do.

Step 5: Determine the scope and format. The SWP can take the form of a flow chart, a hierarchical checklist or both. The appropriate format, which may involve using a template, depends on the complexity of the construction activity.

Step 6: Outline the SWP. List, in writing and in chronological order, the exact tasks that must be carried out for the procedure to be successful.

Step 7: Review, test and refine the SWP. Test the SWP as documented by carrying out the procedure, doing all the tasks in the prescribed order and identify what can be improved. Include control points.

Step 8: Implement the SWP. Once satisfied that the SWP has been approved, it must be implemented and monitored in practice on the construction site.

Note that most well-written SWPs are:

- easy to understand and written with simple, clear instructions in a consistent style, format and level of detail
- brief and include only the most relevant details and, where necessary, references can be included with supporting materials
- actionable, as the construction supervisors and workers should know exactly what actions to take and in what order to complete the specific construction activity
- accessible so that all those who need the SWP can find them easily.

4. Format of a SWP/SOP

Many organisations – and construction firms are no exception – have developed their own unique templates for SWPs because they realise that documenting and mapping procedures are important. There is no one-size-fits-all option; however, consistency is key. Well-written SWPs are action-oriented and should be solid, precise, factual, concise and to the point. They should be available for all tasks that entail risk.

Some organisations even have an SWP for writing SWPs. When everyone involved has clear-cut, comprehensive SWPs that are easy to follow, they will always have a reliable documented guide to help them know precisely what to do. The SWP ensures that, irrespective of who executes a hazardous construction activity, it will be done in a consistent prescribed manner. As such, they can be used in training, audits and induction.

The following is an example of an SWP template:

Example 19.1: An SWP template

ABC CONTRACTORS
SAFE WORKING PROCEDURE (SWP) TEMPLATE

Version:

Approved by:

Date:

Purpose

Include a simple statement regarding why you are writing this SWP. It may also be helpful to describe the purpose of the content in the SWP.

Scope

Describes to whom or under what circumstances the SWP applies.

Definitions/acronyms

First term: Definition of first term

Second term: Definition of second term

Procedures

Describe the construction activity in detail and sequential steps using flow charts and/or hierarchical checklists:

- *Step 1:*

  — *Step 1.1:*

  — *Step 1.2:*

- *Step 2:*

Related resources

Include references and links to relevant materials which may be helpful for executing the construction activity and SWP.

Table 19.1 contains a list of construction activities that can be considered to be high risk and hazardous. It is advisable that SWPs should be developed for each of these.

Table 19.1: Hazardous and high risk construction activities

Construction work where there is a risk of a person falling more than 2 metres
Construction work on telecommunications towers and other similar structures (refer to Construction Regulations, 2014 definition of structures)
Construction work involving demolition
Construction work involving the disturbance or removal of asbestos because special mandatory procedures must be followed
Construction work involving structural alterations that require temporary support such as bracing to prevent collapse
Construction work involving a confined space
Construction work involving trench and bulk excavation to a depth greater than 1.5 metres
The construction of tunnels
Construction work involving the use of explosives
Construction work on or near pressurised gas distribution mains and consumer piping
Construction work on or near chemical, fuel or refrigerant lines
Construction work on or near energised electrical installations and services
Construction work in an area that may have a contaminated or flammable atmosphere, especially below ground level
Tilt-up and precast concrete construction work
Construction work on or in close proximity to roadways or railways used by road or rail traffic
Work on construction sites where there is any movement of powered heavy duty mobile plant and equipment
Work on construction sites where there are artificial extremes of temperature
Construction work in, over or adjacent to water or other liquids where there is a risk of drowning
Construction work involving diving

The following two examples of SWPs were developed by the South Australian Department of Education and may be downloaded from their website (https://www. education.sa.gov.au).

Example 19.2: Example of a SWP for the use of a radial arm saw

(YOUR BUSINESS NAME HERE) – SAFE WORKING PROCEDURE

RADIAL ARM SAW

DO NOT use this machine unless you have been instructed in its safe use and operation, and have been given permission to do so.

PERSONAL PROTECTIVE EQUIPMENT

 Safety glasses must be worn at all times in work areas.	 Long and loose hair must be contained.	 Hearing protection must be worn.
 Sturdy footwear must be worn at all times in work areas.	 Close fitting/protective clothing must be worn.	 Rings and jewellery must not be worn.

PRE-OPERATIONAL HEALTH AND SAFETY CHECKS
- ✓ Locate and ensure you are familiar with all the operations and controls of the radial arm saw.
- ✓ Ensure all guards are fitted, secure and functional. Do not operate the radial arm saw if the guards are missing or faulty.
- ✓ Check the work area and walkways to ensure no slip/trip hazards are present.
- ✓ Keep the work table and work area clear of all tools, off-cut timber and sawdust.
- ✓ Start the dust extraction unit before using the radial arm saw.

OPERATIONAL HEALTH AND SAFETY CHECKS
- ✓ Keep your hands away from the blade and cutting area.
- ✓ The workpiece to be cut must be securely held against a fence.
- ✓ Allow the radial arm saw blade to obtain maximum speed before making a cut.
- ✓ Operate the radial arm saw with the left hand where possible.
- ✓ Avoid reaching over the saw line. Do not cross your arms when cutting.
- ✓ When pulling the radial arm saw across with your right hand, keep your left hand, especially the thumb, well clear of the line of cut.
- ✓ Return the cutting head to the rear of the work table after each cross cut.
- ✓ When cutting bowed timber, place the bow against the work table to avoid the radial arm saw binding.
- ✓ Before making adjustments, switch off the radial arm saw and bring it to a complete standstill.

ENDING OPERATIONS AND CLEANING UP
- ✓ Switch off the radial arm saw.
- ✓ Reset all guards to a fully closed position.
- ✓ Leave the radial arm saw in a safe, clean and tidy state.

POTENTIAL HAZARDS
- ⓘ The radial arm saw may grab and 'kick-back' towards the saw operator.
- ⓘ Flying chips and airborne dust.
- ⓘ Contact with rotating blade.
- ⓘ Eye injuries.
- ⓘ Noise.

→

DO NOT

- ✖   Do not use a faulty radial arm saw. Immediately report to a supervisor or a foreman a suspect radial arm saw.
- ✖   Do not cut branches, dowels or wood with embedded nails or screws.
- ✖   Do not rip solid timber along the grain.
- ✖   Do not cut short lengths of timber.
- ✖   Do not exceed the maximum cut for the radial arm saw.

This SWP does not necessarily cover all possible hazards associated with the radial arm saw and should be used in conjunction with other references. It is designed as a guide to be used to compliment training and as a reminder to users prior to using the radial arm saw.

Example 19.3: Example of a SWP for the use of a forklift truck

(YOUR BUSINESS NAME HERE) – SAFE WORKING PROCEDURE

FORKLIFT TRUCK

DO NOT use this equipment unless you have been instructed in its safe use and operation and have been given permission to do so.

PERSONAL PROTECTIVE EQUIPMENT

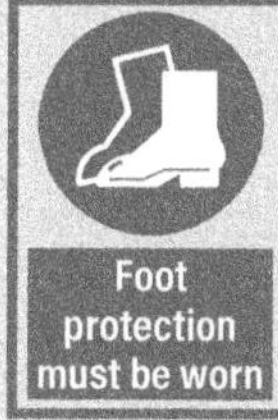

PREOPERATIONAL HEALTH AND SAFETY CHECKS

- ✓   Locate and ensure you are familiar with all the operations and controls of the forklift truck.
- ✓   Check the brakes, lights and hooter of the forklift truck before using it.
- ✓   Ensure that the reversing beeper and warning lights are fully operational.
- ✓   Ensure that the seat belt/safety restraint is in a good condition.
- ✓   Know the capacity of the forklift before using it. Do not use a forklift without a load rating plate.
- ✓   Ensure that the lifting tines are in sound condition and centred either side of the mast.
- ✓   Check all the tyre pressures. Never drive with a flat or under-inflated tyre.

OPERATIONAL HEALTH AND SAFETY CHECKS

✓ Watch out for pedestrians.

✓ Ensure that the lifting tines are secure into the pallet and that the load is stable before lifting or driving off.

✓ Be careful of the ceiling clearances or overhead obstructions when raising the mast of the forklift truck.

✓ Always remember that the safe working load of a forklift reduces as the mast is tilted forward.

✓ Always put the heavy end of the load against the load backrest.

✓ When approaching a blind corner, use the hooter and drive slowly.

✓ Always have someone else guide you if a load restricts your vision.

✓ Slow down when changing direction or when driving on wet or greasy surfaces.

✓ Avoid harsh braking, especially when carrying a load.

REFUELLING

✓ Refuel petrol or diesel engine-powered forklift trucks and recharge battery-operated forklift trucks in a safe and well-ventilated area.

ENDING OPERATIONS AND CLEANING UP

✓ When stopping the forklift:
  – Park on even ground and lower the forks to the ground.
  – Shift the gear selector to the park position and apply the parking brake.
  – Turn off the ignition and remove the keys.

DO NOT

✗ Do not use a faulty forklift truck. Report any faults immediately to a supervisor or a foreman.

✗ Do not use an engine powered forklift truck in poorly ventilated areas.

✗ Do not allow anyone to ride on the tines of the forklift truck.

✗ Do not lift a load with the forklift truck's mast tilted forward.

✗ Do not travel with tines raised or reach mechanism extended.

✗ Never travel with the load elevated.

✗ Do not attempt to turn on an incline or a sloping surface.

✗ Do not dismount from a forklift truck while the engine is running, unless the vehicle has completely stopped, the transmission is in the park position and the parking brake is effectively engaged.

✗ Do not leave tines elevated when the forklift truck is unattended.

✗ Do not refuel an engine-powered forklift truck unless the motor is stopped and ignition turned off.

This SWP does not necessarily cover all possible hazards associated with the use of a forklift truck and should be used in conjunction with other references. It is designed as a guide to be used to compliment training and as a reminder to users prior to using a forklift truck.

5. Review Questions

- Who should be responsible for the development of an SWP?
- When should SWPs be developed?
- What should be included in an SWP?
- Describe the steps to be taken to produce an SWP?

6. Review Exercises

- Using the template shown in the previous example, develop a SWP for the erection of scaffolding on a high-rise building between two floors.
- Similarly, develop a SWP for the use of a paving breaker.

Bibliography

Goetsch, DL. 2013. *Construction safety and health*. Upper Saddle River: Pearson Education, Inc.

Griffith, A & Howarth, T. 2014. *Construction health and safety management*. Harlow: Pearson Education, Inc.

Health and Safety Commission. 2007. *Managing health and safety in construction: Construction (Design and Management) Regulations 2007: Approved code of practice*. Sudbury: Health and Safety Executive Books.

Hinze, J. 2006. *Construction safety*. Gainesville, Fl: Jimmie Hinze.

Holt, ASJ. 2005. *Principles of construction safety*. Osney Mead, Oxford, UK: Blackwell Science Ltd.

National Association of Home Builders. 2007. *Home builders' safety program* Washington: Builderbooks.

Workplace Health and Safety Queensland. 2015. *Masonry wall safety during construction work*. Brisbane: Queensland Government, https://www.worksafe.qld.gov.au/__data/assets/pdf_file/0024/16638/masonry_wall_safety.pdf.

Construction Health and Safety Toolbox Talk Outline

Theo C Haupt

1. Introduction

The construction industry is one that is constantly changing, with change on any construction site being accompanied by many different hazards. The exposure to these hazards must be effectively mitigated against to ensure a safe and healthy working environment on construction sites. Toolbox talks are useful to guide construction workers on site on how to deal with the hazards that they are exposed to on a construction project site. Research has shown that having relevant and regular toolbox talks in the morning before work begins is an effective strategy to reduce the risk of construction workers being injured on the construction site.

2. What is a Toolbox Talk?

A construction toolbox talk is a short health and safety (H&S) message that is used to address hazards on construction sites, share better practices for mitigating and managing hazards exposures and reinforcing H&S requirements before construction workers start their working day or shift on the construction site. In fact, it is often said that the term 'toolbox talk' originates from the image of construction workers standing around their toolboxes in the morning before work, preparing for the day ahead.

Usually, a supervisor or H&S representative is responsible for choosing a relevant construction H&S topic to present to the construction workforce on site. These toolbox talks take place on the construction site itself and can be as short as five or ten minutes. Toolbox talks that form an integral part of the construction H&S plan and management system are an effective way to consistently address the health and safety of construction workers on the site.

3. Frequency of Toolbox Talks

While there are no prescribed guidelines or regulatory requirements with regard to how often toolbox talks should presented, research has shown that daily or weekly toolbox talks are more effective than having lengthy planned or unplanned intervals of time between toolbox talks. Many companies with excellent construction H&S performance records often have toolbox talks before the start of every work shift on the construction project. There are also many construction companies that choose to hold weekly scheduled toolbox talks. When it is considered that taking five or ten minutes out of each working day of the week, having a toolbox talk adds up to more than 1 200 to 2 400 minutes (20 to 40 hours) worth of construction H&S training annually for every construction worker in the organisation.

4. Documenting Construction H&S Toolbox Talks

Effective construction H&S management requires that all H&S training should be written up and documented. The most common way to track attendance at and participation in toolbox talks is by the use of mandatory sign in sheets, which should include the date, time, location on the construction site, construction H&S topics discussed, any discussion notes and the names of the presenter(s). Example 20.1 is an example of a typical toolbox talk sign in sheet.

Example 20.1: Example of construction H&S toolbox talk sign in sheet

Construction Health and Safety Toolbox Talk Sign In Sheet	
Date:	Speaker(s)/Presenter(s):
Topic:	Companies:
Location:	
Additional comments:	
Attendee name	Attendee signature

Every construction worker who attends a toolbox talk must complete and sign the sign in sheet. These sign in sheets, duly completed and signed, should be kept in the construction project H&S file.

5. Format of a Construction H&S Toolbox Talk

As with most aspects of construction H&S documentation, there is no prescribed format for construction H&S toolbox talks in the legislative or regulatory framework. The Construction Regulations 2014 and the Occupational Health and Safety Act, 1993, as amended, include a few documents in prescribed forms sections, but these do not generally relate to the management of construction H&S on construction sites. Instead they relate to various applications and to medical records.

The construction H&S topic for a toolbox talk should be relevant to the construction activities that are currently being undertaken or in the process of being completed on the construction site. The selected topic should be related to specific examples of what is taking place on the construction site. In this way, the construction workers are drawn in and find relevance in the information that is being shared.

It is good common practice and is advisable to compile a construction H&S toolbox talk as an outline. However, the presentation of the information may take many forms using various media. The use of visual templates, illustrations, real tools, plant and equipment from the site and role models, if possible, add value to relaying the information and capturing and holding the attention of the workers. The presentation itself should be kept informal and conversational, yet respectful. The presentation can be made interactive by encouraging the construction workers attending to participate by asking questions about the H&S topic. The presenter could also ask the construction workers leading questions to encourage their participation. The workers could also be asked to share their own experiences relating to the topic of the toolbox talk. It is important that the presenters does a wrap up or summary of the toolbox talk at the end of the talk to reinforce the important points.

Example 20.2 provides an outline of a construction toolbox talk on alcohol. The presentation can be improved by including relevant and appropriate pictures, photos and illustrations.

Example 20.2: Outline of toolbox talk on effects of using alcohol on construction sites

SLIDE NUMBER	CONTENT OF SLIDE
1	Introduction Alcohol and construction work are not compatible
2	Alcohol related problems on the construction site can include: ■ Loss of productivity and poor performance ■ Lateness and absenteeism ■ Increased risk of injuries, incidents, accidents and fatalities ■ Effect on morale and construction worker relations ■ Bad behaviour or poor discipline ■ Adverse effects on the company image and client relations Alcohol is a depressant drug, which depresses parts of the brain function
3	Importance of the topic: ■ Alcohol is estimated to cause, on average, 3–5% of all absences from work with huge loss of working days each year ■ Alcohol consumption may result in reduced work performance and the resentment of the other construction workers who have to carry the workload of workers whose work output declines because of their drinking ■ Even at blood alcohol concentrations lower than the legal drink/drive limit, alcohol reduces physical coordination and reaction speeds ■ It also affects thinking, judgement and mood
4	Regulations and requirements: ■ Employers have a general duty under the Occupational Health and Safety Act to ensure, as far as is reasonably practicable, the health, safety and welfare of their workers ■ Knowingly allowing a construction worker to continue working under the influence of excess alcohol, and placing that worker and others at risk, could result in prosecution ■ Construction workers are also required to take reasonable care of themselves and others who could be affected by what they do ■ In high-risk industries like the construction industry, there is additional legislation in place to control the misuse of alcohol and drugs
5	Guidance: ■ For many workers, drinking alcohol is a positive part of life and does not cause any problems ■ Drinking too much or at the wrong time can be harmful ■ The misuse of alcohol can lead to reduced productivity, taking time off work and accidents on the construction site ■ Some employers adopt alcohol screening as part of their alcohol policy ■ The good news is that construction workers with drink problems can and do cut down on their alcohol consumption after receiving expert help and counselling ■ There are places throughout the country where workers with drinking problems can go for expert help

→

SLIDE NUMBER	CONTENT OF SLIDE
6	What happens when you drink: ■ Alcohol is absorbed into your bloodstream within a few minutes of being drunk and carried to all parts of your body, including the brain ■ The concentration of alcohol in the body, known as the 'blood alcohol concentration', depends on many factors but, principally, on how much you have drunk, how long you have been drinking, whether you have eaten and your size and weight ■ It is difficult to know exactly how much alcohol is in your bloodstream or what effect it may have on you ■ It takes a healthy liver about 1 hour to break down and remove 1 unit of alcohol ■ If someone drinks two pints of ordinary strength beer at lunchtime or half a bottle of wine, they will still have alcohol in their bloodstream three hours later ■ If someone drinks heavily in the evening they may still be over the legal drink/drive limit the following morning ■ Black coffee, cold showers and fresh air do not sober someone up ■ Only time can remove alcohol from the bloodstream
7	Don't drink at all: ■ Before or while driving ■ Before using any tools, machinery, electrical equipment or ladders ■ Before working on the construction site, when appropriate functioning would be adversely affected by alcohol
8	Question 1: How many units does a healthy liver remove in one hour? a. One b. Two c. Three d. Five
9	Question 2: What can help remove alcohol from the blood stream? a. Black coffee b. Cold showers c. Fresh air d. Time

Example 20.3 is an example of an outline of a toolbox talk on asbestos. The presentation can be improved by including relevant and appropriate pictures, photographs and illustrations.

Example 20.3: Outline of toolbox talk on exposure to asbestos

SLIDE NUMBER	CONTENT OF SLIDE
1	Introduction: ■ This talk will cover where you might find asbestos, how it can affect you and hazardous work ■ Asbestos was used extensively in building materials from around the 1950s and was eventually banned around the world 50 years later ■ Any building built or refurbished within that period is likely to contain asbestos ■ Asbestos fibres are so tiny, they cannot be seen by the naked eye
2	Importance of topic: ■ Asbestos fibres can kill ■ Around 4 500 people a year, from all industries, die from asbestos-related diseases ■ Each week, 40 artisans on average die as a result of asbestos exposure at work
3	Regulations and requirements: ■ Clauses (2) to (5) of the Asbestos Regulations, 2001 place a number of requirements on employers in an attempt prevent reduce the risk of exposure occurring ■ One of these requirements is mandatory training for anyone who is likely to be exposed to asbestos fibres at work
4	Guidance Where asbestos is commonly found: ■ Insulation and sprayed coatings used for boilers; plant and pipe work hidden in under-floor ducting; fire protection applied to steelwork hidden behind false ceilings; thermal and acoustic insulation of buildings; some textured coatings and paints; and packing in heating and ventilation systems ■ Insulating board used in the following places: fire protection to doors, protected exits and steelwork; claddings on walls and ceilings; and internal walls, partitions and suspended ceiling tiles ■ As asbestos cement, which is found in corrugated roofing and cladding sheets of buildings; flat sheets for partitions, cladding and door facings; and rainwater gutters and down pipes
5	How asbestos can affect you: ■ Asbestos breaks into tiny, long, sharp fibres. They can scar the lungs, causing asbestosis or fibrosis ■ Asbestos fibres may also cause lung cancer ■ It can also cause mesothelioma, a cancer of the inner lining of the chest wall. This cancer is incurable ■ Smokers are at much greater risk of asbestos-related diseases
6	Hazardous work: ■ Plumbers, carpenters and electricians working on building repair are considered most at risk. However, all trades can come into contact with asbestos ■ Buildings constructed or refurbished before the year 2000 may contain asbestos materials
7	Top tips: ■ Always check the asbestos survey for any asbestos-containing materials before working on any buildings built prior to the year 2000 ■ Never disturb asbestos-containing materials unless you have been trained to do so and have the appropriate PPE and controls in place

→

SLIDE NUMBER	CONTENT OF SLIDE
8	Question 1: Where is asbestos commonly found? a. Insulation b. Cement-based products c. Ceiling tiles d. All of these
9	Question 2: How can asbestos exposure affect you? a. Asbestos fibres can damage the lungs causing curable respiratory problems b. Asbestos fibres can damage the lungs causing incurable respiratory problems c. Asbestos fibres can act as a barrier to other harmful dusts and fumes d. Asbestos fibres can be harmful to the lungs over prolonged exposure

6. Review Questions

- When should toolbox talks be developed?
- How often should toolbox talks be presented?
- How would you control attendance at toolbox talks?
- What could improve toolbox talks?

7. Review Exercises

- Draft a toolbox talk for using an angle grinder properly and safely
- Draft a toolbox talk on personal protective equipment using the guides available at https://www.haspod.com/blog/toolbox-talks/free-toolbox-talks-construction.

Bibliography

HASprod. 2021. 30+ Free Toolbox Talks For Construction, https://www.haspod.com/blog/toolbox-talks/free-toolbox-talks-construction.

Health and Safety Executive. nd. Toolbox talks, https://www.hse.gov.uk/construction/resources/toolboxtalks.htm .

Hinze, J. 2006. *Construction Safety.* Gainesville: Jimmie Hinze.

National Association of Home Builders. 2007. *Home builders' safety program.* Washington: Builderbooks.

CHAPTER 21

Construction Health and Safety File

Theo C Haupt

1. Introduction

The Construction Regulations, 2014 require the principal contractor (PC) on a construction project to compile a construction health and safety (H&S) file. The notion of a construction H&S file is unfortunate and is probably the most misunderstood aspect of managing construction health and safety. Once the purpose of the construction H&S file is understood, the confusion about what should be included and what should be left out becomes clearer. When compiled properly and correctly, as intended by the Construction Regulations, 2014, it provides a powerful management and information resource to protect everyone involved in current and future projects. Therefore, it should form part of any H&S management system in place for the construction project. It is possible that a construction H&S file already exists, relevant to the current project being undertaken. In these instances, the file must be included in the pre-construction information provided by construction clients to ensure that contractors and designers preparing for the new project can take the information into account.

2. Purpose of the Construction H&S File

The construction H&S file must be appropriate to the characteristics of the construction project. The purpose of the construction H&S file is very clear, it is intended to ensure that any party who may carry out future construction work on a structure or site, such as, for example, cleaning, alteration, refurbishment, repairs, maintenance, construction or demolition, are made aware of the significant construction H&S risks that may be encountered. Therefore, the construction H&S file is for use on future projects and only information that serves that purpose should be included in it. In this way, the file is an important tool to enable a party to plan and develop healthy and safe systems of work, and manage their construction work without risk to their own health and safety and the health and safety of others who may be affected by their construction activities. The construction H&S file will provide information that is essential to those responsible for doing the construction work. As such, it alerts them to the risks involved and helps them to decide on how to execute the construction activities safely. The PC is responsible for keeping the construction H&S file appropriately reviewed, updated and revised from time to time to take account of the construction work and any changes that may occur during the project. Other members of the construction project team must provide relevant information to the PC as the construction work proceeds so that the file can be compiled and always kept current. At the end of the project, the PC must pass the construction H&S file onto the client for future use.

3. Contents of the Construction H&S File

Given the purpose and intention of the construction health and safety file, the following are suggestions for what should be included in it:

- a brief description of the construction work carried out
- any hazards that have not been eliminated through the design and construction processes, and how they have been addressed on the project, such as, for example, surveys or information concerning asbestos or lead
- key structural principles used in the constructed structure, such as, for example, pre- or post-tensioned members and the safe working loads for the various floors and roofs
- hazardous materials used, including epoxy coatings and other special coatings
- information regarding the removal or dismantling of installed construction plant and equipment, such as any special arrangements for lifting such equipment
- H&S information about any equipment provided for cleaning or maintaining the structure
- the nature, location and markings of significant services, such as any underground cables, gas supply equipment, electricity main supplies, air conditioning ducts, water supply lines and firefighting services
- information and as-built drawings of the building, its plant and equipment that include details of the means of safe access to and from service voids and fire doors.

On the other hand, the construction H&S file should not include any information or details that will not help with planning for construction H&S in respect of future construction work on the completed structure, such as, for example:

- the preconstruction information or a construction phase plan
- construction phase risk assessments, safe working procedures and other written systems of work and that were adopted during the project

- details about the normal operation and use of the structure after its completion
- construction phase accident statistics
- details of all of the contractors and designers involved in the project although, arguably, it may be useful to include details of the PC, design engineers and specialist services and materials suppliers and manufacturers
- contractual documents
- information about structures, or parts of structures, that have been demolished during the course of construction, unless there are any implications for remaining or future structures
- information contained in other documents that could be referenced and cross-referenced.

4. Management and Control of the Construction H&S File

The construction H&S file should be held on the construction site in a convenient location where it will be available for inspection as the need arises. It can be in paper form or electronic form. Where the construction H&S file is kept electronically, it is important to ensure that suitable back up and storage arrangements are put in place. The level of detail of the construction H&S file must be in proportion to the risks involved on the construction project, while the degree of effort and resources expended to compile it must also be proportionate to the risks involved on the construction project.

The construction H&S file must always be kept current and up to date, and retained for the entire lifecycle of the structure. Where clients dispose of their entire interest in a particular structure, they should hand the construction H&S file over to the new owners. Clients should ensure that the new owners are aware of the nature and purpose of the file. Where they sell part of a structure, any relevant information contained in the construction H&S file pertaining to that part of the structure should be provided or copied to the new owner.

5. Guidelines for Compiling the Construction H&S File

The guidelines contained in Table 21.1 will assist the PC to compile a satisfactory construction H&S file. Explanations are provided for each suggested section of the file.

Table 21.1: Guidelines to compile a construction H&S file

1.	The Project	Introduction, Construction Statement and Project Directory
		Project Description What was the purpose of the project? What was built, installed or demolished? A brief description at the start of the construction H&S file will assist interested parties in the future to know what the construction H&S file covers, and whether the file is relevant to them. This section should include: ■ description of the works ■ location of the site ■ project team directory ■ subcontractor register ■ supplier register ■ key dates, such as the project start date and project completion date.
2.	Residual Hazards	Statements of Residual Hazards
		Residual hazards Are there any residual hazards on the construction site that may affect parties involved in the future use, maintenance, cleaning or demolition of the work? For example, were any hazardous materials left in place? Was asbestos left undisturbed? Did the ground investigation highlight any issues? Examples include: ■ ground conditions ■ asbestos ■ fragile materials ■ contamination ■ other residual hazards.
3.	Hazardous Materials	Schedule of Unusual Hazardous Materials
		Hazardous materials It is not just old hazards that may remain on the construction site that future users need to be aware of. What about the materials that have been installed/used during the construction of the structure? Future building users might not know from looking at a material that it is hazardous but the principal contractor will know all the about the materials that have been used and passing the information on could prevent ill health or a fatality in the future. Is it necessary to provide H&S information for the following, for example: ■ paints ■ coatings ■ hazardous substances ■ flammable substances.
4.	Key Structural Principles	Description of Structural Design and Demolition Statement
		Has the completed construction project changed an existing structure or created a new structure? The client and future building users and maintainers need to know about any adjustments and limitations of the structure. These could include details of, for example: ■ safe working loads ■ structural design ■ structural alterations.

$\rightarrow$

5.	Plant and Equipment	**Principles of Services Design, the Cleaning and Maintenance Access Strategy, and a Plant Replacement/Removal Strategy**
		Safe removal In the same way that construction H&S was managed during the construction phase, the same must be done when removing and dismantling the structure, the plant and equipment inside it. Installed equipment will be replaced and removed and, eventually, the building might be too. ■ decommissioning ■ dismantling instructions ■ lifting arrangements.
6.	Building Fabric	**Description of Architectural Design, the Cleaning and Maintenance Access Strategy for the Structure, and a Schedule of Access Equipment**
		H&S-specific operation and maintenance information should be included, such as, for example: ■ safe methods ■ safe access ■ temporary access ■ safe cleaning.
7.	Significant Services	**Known Underground Services and a Description of Emergency and Firefighting Systems**
		Location of services Services are a construction H&S concern on any project because they are often hidden from view. They can be in walls, under floors and underground. The risks to future building and site owners can be reduced by providing this information about the completed project. The principal contractor will know the locations of these installed services that include, for example, where cables have been laid, where access points are and where shut-off valves are located. The services include the following: ■ cables ■ ducts ■ gas ■ electrical ■ water ■ underground services.
8.	As-Built Information	**As-Built Drawing Register and a Schedule of Operation and Maintenance Manuals**
		As-built The final plans for the construction project show what has been built, where plant and equipment have been installed, and where access to voids, shafts and other serviceable parts of the building are. The following should be included: ■ as-built drawings ■ as-installed drawings.
9.	Surveys and Reports	**Relevant Surveys and Reports**
10.	Other information	**Maintenance Manuals and Operation Information not Related to H&S**

The more organised and relevant the construction H&S file is, the better it will be and the more useful it will be for helping future work to be carried out safely, without any threat to those involved in the future project. Including too much irrelevant material may cause crucial information about risks on the future project to be overlooked. Table 21.2 is an example of a suggested table of contents for a construction H&S file.

Table 21.2: Table of contents of a construction H&S file

TABLE OF CONTENTS

Guidance note:
1.0 Background
    1.1 Site description
    1.2 Project description
2.0 Design criteria
    2.1 Element/scope of work
    2.2 Design, criteria, standards and assumptions made
    2.3 Non-standard or unusual features of the design
    2.4 Relevant documents, such as surveys, design reports and calculations
3.0 Residual hazards
4.0 Removal, dismantling or demolition
5.0 Maintenance facilities and procedures
6.0 Operating and maintenance manuals
7.0 As-built drawings and key documents
8.0 As-installed drawings and key documents
9.0 Other information
10.0 Document control

6. Review Questions

- Who should be responsible for the compilation of the construction H&S file?
- What is the purpose of the construction H&S file?
- What should be included in the construction H&S file?
- What should not be included in the construction H&S file?

7. Review Exercise

Compile a construction H&S file for the construction of a two-storey house constructed of loadbearing plastered and painted brickwork on concrete strip footings; with a pitched tiled roof with PVC rainwater system; aluminium windows and external doors; skimmed gypsum rhino board ceilings; ablution areas tiled floor to ceiling with tiled dados over kitchen sink; tiled floors in kitchens and bathrooms and timber laminated flooring throughout the rest of the house; standard electrical, plumbing and drainage fittings and appliances; with electrified boundary fencing.

Bibliography

Griffith, A & Howarth, T. 2014. *Construction health and safety management*. Harlow: Pearson Education, Inc.

Health and Safety Commission. 2007. *Managing health and safety in construction: Construction (Design and Management) Regulations 2007: Approved code of practice.* Sudbury: Health and Safety Executive Books.

Holt, ASJ. 2005. *Principles of construction safety*. Osney Mead, Oxford: Blackwell Science Ltd.

CHAPTER 22

Construction Health and Safety Inspections and Audits

Theo C Haupt

1. Introduction

All construction employers are morally, ethically and legally obliged to provide their construction workers with safe and healthy working conditions on the construction sites. To know how unsafe and health-threatening working conditions are or could become on the construction site, employers must constantly monitor performance against health and safety (H&S) standards and procedures, as prescribed by law and their own policies. This knowledge can be achieved by conducting regular H&S inspections on the construction site and recording the results. In fact, construction H&S inspections can be a vital part of injury and accident prevention efforts, if done well. Inspections can help reassure workers that the workplace is safe, with no threat to their health, and help the construction organisation demonstrate that it cares about the well-being of its workers and stakeholders. In order for this to happen, a robust process is necessary. Construction H&S inspections help prevent incidents, injuries and illnesses. By critically examining the construction site, inspections help to identify and record hazards for corrective action. Construction H&S committees can help plan, conduct, report and monitor inspections. Regular construction H&S inspections are an important part of the overall construction H&S programme and management system.

2. Considerations

As tools and equipment used on construction projects and sites get older, they become worn out or damaged. This equipment can become very dangerous to use on the site if not repaired or replaced. Construction materials, equipment and procedures constantly change and ignoring such changes can jeopardise the health and safety of construction workers. A lack of the necessary skills to perform a certain construction activity can cause injury. Regular construction H&S inspections will help to expose possible training needs so that these activities can be done safely.

H&S inspections may vary in purpose, from routine maintenance inspections to an inspection after a major incident, like a fire on the construction site, for example. Planned construction H&S inspections, sometimes referred to as 'walkabouts', are usually planned based on accident statistics; the presence of hazardous substances, such as the use of chemical additives in concrete; or the use of dangerous heavy duty plant and equipment on the construction site. Unplanned inspections, on the other hand, usually arise from requests or complaints by construction workers, employers or members of the public.

In terms of the Construction Regulations, 2014, clients may mandate designers or their construction H&S agent to conduct regular inspections. Similarly, contractors must conduct regular inspections on their construction sites as the work proceeds, to ensure compliance with the project construction H&S plan.

3. Purpose of Construction H&S Inspections

Construction H&S inspections are important because they allow construction employers to:

- listen to the concerns of their construction workers and supervisors
- gain a further understanding of construction activities and tasks
- determine the underlying causes of hazards on construction sites
- identify potential hazards on the site before they become a serious problem, with the aim of removing or controlling them
- act on the findings of inspection, which serve as an early warning system
- check whether relevant construction H&S standards and laws are being adhered to
- check whether any agreed improvements to construction activities are carried out on the construction site
- establish the cause of injuries, death, occupational diseases, property damage and pollution as a result of construction activities
- record all necessary construction H&S information and report to the appropriate authorities in the required format
- provide information at various meetings where construction H&S matters are discussed
- make information available for relevant research purposes
- assist in the evaluation of the construction H&S policy and procedures, including the completion of all the registers, records and schedules required by law.

From the information contained in the reports of the construction H&S inspections, the construction employer will be able to formulate corrective and/or remedial action. The employer will be able to monitor steps taken to eliminate hazards or control risks, for example as by considering substitutes; introducing engineering and administrative controls, policies, and procedures; and proper use and maintenance of personal protective equipment (PPE).

4. Common Construction Site Hazards

Common construction site hazards include:
- safety hazards such as those caused by inadequate machine guards, unsafe construction site conditions and unsafe execution of construction activities
- biological hazards caused by organisms such as viruses, bacteria, fungi and parasites
- chemical hazards caused by a solid, liquid, vapour, gas, dust, fume or mist
- ergonomic hazards caused by the physiological and psychological demands on the construction worker, such as repetitive and forceful movements, awkward postures arising from improper methods of executing construction activities and improperly designed construction workstations, tools, and equipment
- physical hazards caused by noise, vibration, energy, weather, heat, cold, electricity, radiation and pressure as a result of construction activities
- psychosocial hazards that can affect the mental health or well-being of construction workers such as overwork, stress, bullying, Covid-19 protocols or violence.

5. Common Poor Construction H&S Practices

Some common poor construction H&S practices on construction sites include:
- using machinery or tools without authority
- operating at unsafe speeds or in other violation of safe working practices or procedures
- removing guards or other safety devices, or making the devices ineffective
- using defective tools or equipment, or using tools or equipment in unsafe and unhealthy ways
- using hands or body instead of tools or push sticks
- overloading, crowding or failing to balance materials, or handling materials in unsafe and unhealthy ways, including improper lifting
- repairing or adjusting equipment that is in motion, under pressure or electrically charged
- failing to use, maintain or improperly use PPE or H&S devices
- creating unsafe, unsanitary or unhealthy conditions by improper personal hygiene, by using compressed air for cleaning clothes, by poor housekeeping or by smoking in unauthorised areas
- standing or working under suspended loads, scaffolds, shafts or open hatches
- discussion with or observation of workers who may be overloaded, fatigued, working in conflict with others, working in isolation or working alone.

6. Effective Construction H&S Inspections

Construction H&S inspections can be improved in many different ways, including some of the following considerations:

- Every construction H&S inspection on the site must examine the who, what, where, when and how. It is important that particular attention is given to items and issues that are or are most likely to develop into unsafe or unhealthy conditions. This development results from stress, wear, impact, vibration, heat, corrosion, chemical reaction, misuse or any combination of these. Areas must be included where no construction activities are done regularly, such as parking areas, rest areas, stores, ablutions and locker rooms, where provided.
- All construction workplace elements must be included and inspected, namely the construction workers, the construction site environment, the construction plant and equipment and the construction activities.
- Frequent construction H&S inspections by themselves may have very little impact on improving the construction H&S culture on the construction site.

7. Resources for Construction H&S Inspections

Construction H&S inspections should make use of the following resources:
- a plan of all areas of the construction site
- a construction health and safety checklist but be careful to avoid tick fever
- materials safety data sheets for all materials used on the site
- construction H&S inspection sheets that will be used during the actual inspection of the site
- information on chemicals and other hazardous substances used on the site
- storage areas
- displayed construction site rules and regulations
- proximity of safe working procedures (SWPs) to hazardous construction activities
- proper use and maintenance of PPE used on the site
- engineering controls used on the site
- emergency procedures for fire, first aid and rescue
- construction worker complaint reports regarding particular hazards in the workplace, including psychosocial hazards
- recommendations of the H&S committee
- previous construction H&S inspection reports and records
- maintenance reports, procedures and schedules for critical plant and equipment
- monitoring reports (levels of chemical, physical or biological hazards)
- reports of unusual operating conditions
- names of inspection team members and any technical experts assisting in the inspections.

Example 22.1 is an example of a construction activity inspection checklist.

Example 22.1: Sample of a construction activity inspection checklist

Work Area: _______________________________

Date of Inspection: _______________________

Person/s Inspecting: ______________________

Criteria	Rating C = Commendable S = Satisfactory U = Unacceptable N = Critical	Actions Required	By Whom	By When	Action Completed
Corridors / Stairs					
No blind corner					
Hand rails accessible					
Anti-slip tread on stairs					
Stairs in good condition					
Floors					
Even, visible steps etc					
In good condition (no trip hazards)					
No spills					
Work areas					
Clean and tidy					
Equipment, paperwork put away					
Storage					
Items stored correctly					
Storage designated to minimise lifting problems					
Walking area clear					
Electrical					
Residual current devices fitted for moveable equipment					
Equipment checked and has current inspection tags					
No cords on floor or across walkways					

Criteria	Rating	Actions Required	By Whom	By When	Action Completed
Equipment					
In good condition (if unsafe, taken out of service)					
In use or stored appropriately					
Suitable for purpose used					
Maintenance checks/ records up to date					
Ventilation					
Air vents, filters and extraction fans clean					
Servicing records kept up to date					
Gas Cylinders					
Cylinders secured					
Stored outside (minimum number inside)					
Cylinders in use secured on trolley					
Hazardous Substances					
Material safety data sheets available for all substances					
All containers clearly labelled					
Stored appropriately					
Manual Handling					
Unnecessary manual handling eliminated					
Staff trained in manual handling					
Lighting					
Light fittings clean and working					
Work areas well lit					
Night lighting adequate					
Security lights working					

$\rightarrow$

Criteria	Rating	Actions Required	By Whom	By When	Action Completed
Welfare facilities					
Toilets, hand basis clean, with soap available					
Lunch room/canteen clean					
Safety Signs					
Construction health and safety policy displayed					
First aid, protective and fire equipment, signs etc posted					
Waste Disposal					
Bins regularly emptied					
Clean					
Food scraps in vermin-proof bins					
Infectious waste disposed of appropriately (if applicable)					
Fire and Emergencies					
Extinguishers in place					
Serviced and not blocked					
Exits clearly marked and clear					
Exits and emergency lighting work					
Action cards and emergency numbers displayed					
Smoke detectors tested					
Fire blankets accessible					
Workers know procedures (ask a sample of staff)					
First aid kit available, well stocked and clean					
Records kept of first aid provided					

The following construction H&S inspection rating system should be considered:

- **Commendable** if the construction H&S policies and practices meet requirements. This would be indicated by evidence that:
  - all criteria are satisfied
  - there is evidence of continuous improvement
  - there is no areas of major H&S risks nor major concerns about the well-being of construction workers/visitors
  - there is a consistent level of achievement of the outcome.
- **Satisfactory** if the construction H&S policies and practices generally meet the requirements and there are only minor deficiencies, which can be rectified within an agreed timeframe. There would be evidence of:
  - an adequate corrective plan currently being implemented to rectify the deficiency within a reasonable time
  - no major areas of health or safety risk or major concern about the well-being of construction workers/visitors.
- **Unacceptable** if any major health or safety risks or concerns about the well-being of construction workers or visitors exist but the organisation can provide evidence that an adequate corrective plan is being actioned and will be achieved within an acceptable timeframe; or if the policies and practices of the organisation do not meet the expected outcome because there are major deficiencies other than a major health or safety risk or concern about the well-being of construction workers which, despite corrective action, will take considerable time to rectify.
- **Critical** if a major health or safety risk or concern about the well-being of construction workers or visitors exists for which there is no evidence of an adequate plan for corrective action currently being implemented. Immediate corrective action is required. **Critical** means no corrective plan is in place and immediate corrective action is required.

An alternative rating approach could be to use a prioritisation approach by assigning a priority level to the hazards observed to indicate the urgency of the corrective action required. For example:

A = Major – requires immediate action

B = Serious – requires short-term action

C = Minor – requires long-term action

8. Principles of Construction H&S Inspections

When conducting H&S inspections, it is useful to follow these basic principles as suggested by the Canadian Centre for Occupational Health and Safety, namely:

- Draw attention to the presence of any immediate danger, other items can await the final report.

- Shut down and 'lock out' any hazardous items that cannot be brought to a healthy and safe operating standard until repaired.
- Do not operate equipment. Ask the operator for a demonstration. If the operator of any piece of equipment does not know what dangers may be present, this is cause for concern. Never ignore any item because of not having the knowledge to make an accurate judgement of health and safety.
- Look up, down, around and inside. Be methodical and thorough. Do not spoil the inspection with a 'once-over-lightly' approach.
- Clearly describe each hazard and its exact location in the rough notes of the inspection. Allow 'on-the-spot' recording of all findings before they are forgotten. Record what has or has not been examined in case the inspection is interrupted.
- Ask questions, but do not unnecessarily disrupt work activities. This interruption may interfere with an efficient assessment of the job function and may also create a potentially hazardous situation.
- Consider the static or stop position and dynamic or in motion conditions of the item being inspected. If a machine is shut down, consider postponing the inspection until it is functioning again.
- Consider factors such as how the work is organised or the pace of work and how these factors impact safety.
- Discuss as a group, 'Can any problem, hazard or accident generate from this situation when looking at the equipment, the process or the environment?' Determine what corrections or controls are appropriate.
- Do not try to detect all hazards simply by relying on senses or by looking at them during the inspection. Equipment may have to be monitored to measure the levels of exposure to chemicals, noise, radiation or biological agents.
- Take a photograph or, if unable to, clearly describe or sketch a particular situation.

9. Types of Construction H&S Inspections

There are several different types of construction H&S inspections. These include:

- **Informal inspections**: Informal inspections are simply the conscious awareness of workers as they go about their regular activities. Properly promoted and utilised, workers may spot many potential problems as changes occur and work progresses. Supervisors and workers continually conduct ongoing inspections as part of their job responsibilities. Such inspections identify hazardous conditions and either correct them immediately or report them for corrective action. The frequency of these inspections varies with the number and condition of equipment use. Daily checks by users ensure that the equipment meets minimum acceptable health and safety requirements.
- **Planned or periodic inspections**: Inspections that are regular, planned inspections of the critical components of equipment or systems that have a high potential

for causing serious injury or illness are referred to as periodic inspections. These inspections are included as part of standard preventive maintenance procedures or hazard control programmes and could be conducted daily, weekly, monthly, annually or at other stipulated interval. Legislation and regulations may require that only suitably qualified or competent persons should be permitted to regularly inspect certain types of equipment, such as lifts, boilers, pressure vessels, scaffolding and fire extinguishers.

- **Critical parts/items inspections**: Critical parts or items may be defined as the components of machinery, equipment, materials, structures or areas more likely than others to result in a major problem or loss when worn, damaged, abused, misused or improperly handled. Effective inspection programmes ensure that all these parts/items are identified, evaluated and kept in proper condition.
- **Housekeeping construction H&S inspections**: Housekeeping inspections are vital as effective planned inspections and include both cleanliness and order. In practice, housekeeping means more than clean and neat; that things are where they ought to be for maximum productivity, quality, safety or cost control.
- **General construction H&S inspections**: A general construction H&S inspection is a planned walk through of an entire area of the construction site or plant. It is comprehensive and inspectors look at anything and everything.

10. Frequency of Construction H&S Inspections

The frequency and duration of inspections will depend on several factors. It is, therefore, recommended that an inspection schedule be developed for each construction site. The inspection schedule should state:

- when to inspect each area or item on the construction site
- who carries out the inspection
- what degree of detail to inspect in each area or item.

11. Composition of a Construction H&S Inspection Team

The construction H&S inspection team should satisfy the following criteria:

- extent of knowledge of construction H&S regulations and procedures
- extent of knowledge of potential hazards and possible mitigating interventions
- extent of actual experience with and knowledge of the construction activities involved.

Example 22.2 is an example of a construction H&S inspection report.

Example 22.2: Sample of a construction H&S inspection report

✓ Inspection Location: _________________________________ Date of Inspection: ______________________________

✓ Department/Areas Covered: _____________________ Time of Inspection: _______________________________

Observations					For Future Follow-up		
Item and Location	Hazard(s) Observed	Repeat Item Y / N	Priority A/B/C	Recommended Action	Responsible Person	Action Taken	Date

✓ Copies to: _______________________________ Inspected by: _________________________________

Example 22.3 presents an alternative construction health and safety inspection checklist.

Example 22.3: Construction H&S inspection checklist

	YES	NO
Asbestos		
Are any areas containing asbestos identified, marked and an up-to-date record kept?		
Is all the asbestos in good condition/sealed in and monitored?		
Has the risk of exposure to asbestos dust and fibres been assessed and an up-to-date written record kept?		
Are there arrangements in place to inform any contractors about any asbestos presence or locations where it is not known whether or not that area is clear of asbestos?		
Is there a plan for specialist removal of asbestos?		
Chemicals		
Are all containers clearly labelled with contents and hazard warnings, and the precautions to be taken?		
Are there safety data sheets for all chemicals, including cleaning and other materials?		
Is training provided in safe use of chemicals and on what to do in an emergency (spillage, poisoning, splashing etc)?		
Cleanliness		
Are all work surfaces, walls and floors kept tidy and regularly deep cleaned?		
Electrical safety		
Are all electrical equipment, fittings and tools regularly checked and maintained?		
Is access to live high voltage equipment restricted to authorised people only?		

	YES	NO
Fire precautions		
Are there separate storage arrangements for flammable materials?		
Are bins regularly emptied and rubbish safely disposed of?		
Are cigarettes and matches disposed of separately from other rubbish?		
Are clear fire instructions displayed throughout the workplace?		
Have sources of ignition (portable heaters etc) been replaced with safer alternatives?		
Are fire drills carried out regularly and at least once per year?		
Are fire alarms and smoke detectors checked and tested weekly?		
Are the alarms capable of warning workers throughout the building?		
Are there other forms of fire warning for the hearing impaired?		
Are all workers given information, instruction and training on fire risks and precautions, as well as what to do in the event of a fire or fire alarm?		
Is emergency lighting provided and tested regularly?		
Are fire escape routes clearly signed, kept clear and wide enough to prevent a crush, and do they lead quickly and directly to a safe area?		
Are fire doors and exits clearly marked, always kept clear on both sides and never left open, and do they open easily and quickly in the direction of escape and lead quickly to a safe area?		
First aid, accidents and illnesses		
Is there a first aid box and is it fully equipped and accessible to workers?		
Is there a trained first-aider or appointed person on the premises?		
Is it clear who the first-aider(s)/appointed person is/are and how they may be contacted?		
Is a clean and properly equipped first aid room available?		
Are all accidents, near misses and illnesses caused by work reported and recorded in an accident book?		
Is safe social and other distancing of at least 1.5 metres practiced?		
Is screening for Covid-19 done?		
Are workers showing symptoms prevented from accessing the workplace?		
Are workers showing symptoms of infection referred for testing?		
Lighting		
Is the lighting bright enough, especially over workstations?		
Are stairs and corridors etc properly lit?		
Are light bulbs replaced promptly?		
Are windows clean on both sides and free from obstructions?		

$\rightarrow$

	YES	NO
Lifting and manual handling		
Have all workers who are at risk from lifting or moving been trained in manual handling?		
Is mechanical equipment used whenever possible, have workers been trained in its use and is there sufficient space to use it in?		
Where mechanical assistance is not possible, are workers trained in safe lifting techniques and is there sufficient space to use these techniques?		
Are heavy items stored at a convenient or adjustable height to suit the user?		
Is the weight of loads known and clearly marked, and are they small and light enough?		
Are unbalanced, uneven, slippery, sharp or too hot or too cold loads avoided?		
Are loads securely packed to avoid them shifting or spilling?		
Are work surfaces at a comfortable or adjustable height to suit the user and at compatible heights to reduce lifting from one to another?		
Is frequent or prolonged stooping, stretching or reaching above shoulder height or sideways twisting of the body avoided?		
Are lifting and handling needs included in patient/client care plans?		
Do uniforms, PPE and other clothing that is provided allow easy movement?		
Machinery and equipment		
Are all workers trained to use, clean and adjust equipment safely?		
Is all equipment regularly inspected and maintained?		
Is there a procedure for reporting faulty equipment and for taking it out of use until repaired?		
Are all guards in place on machinery?		
Are potentially dangerous machines only operated by properly trained workers aged 18 years old and older?		
Noise		
Are noise levels below the recommended maximum (rough guide – you should be able to talk with someone a metre away without shouting)?		
Have the causes of noise been tackled?		
As a last resort, are suitable earmuffs or plugs provided and are they regularly checked, cleaned, maintained and stored in a clean and safe place?		
Protective clothing		
Is proper and appropriate protective clothing provided free of charge?		
Is it effective, comfortable and well fitting?		
Is it replaced as soon as it are worn out or damaged?		
Are clean overalls provided regularly?		

	YES	NO
Stress		
Do risk assessments include stress?		
Has your employer done a stress audit?		
Are there measures in place to avoid or minimise the risk of stress?		
Has the employer introduced the stress management standard?		
Temperature (working indoors).		
Is the temperature comfortable all year?		
Does the temperature reach at least 16 °C within one hour of starting work?		
Can breaks be taken away from hot areas?		
Temperature (working outdoors)		
Is warm clothing provided in cold weather?		
Are there facilities for warming up and making hot drinks when cold?		
In hot conditions, is cool drinking water provided and can breaks be taken in the shade?		
Can the work be organised so that it takes place in the shade or not during midday when the sun is at its strongest?		
Toilets, wash and rest facilities		
Are there sufficient toilets and are they clean and in good repair?		
Are washing facilities (hot water, soap and towels) provided?		
Are sanitary disposal facilities provided in women's toilets?		
Are lockers (or something similar) provided for workers?		
Is there a rest room and is it clean, properly lit and ventilated?		
Are there suitable facilities for pregnant and nursing mothers to rest?		
Are there facilities for workers to eat meals?		
Ventilation		
Are fumes, steam and stale air removed?		
Is there a supply of fresh air without draughts?		
Are special precautions taken when working in confined spaces?		
Violence		
Has a risk assessment on violence or the threat of violence been conducted (physical, verbal abuse or intimidation)?		
Are workers encouraged to report all incidents of violence, including intimidation?		
Are there preventive measures in place to avoid or minimise the risk of violence?		
Are workers trained in what to do and how to diffuse potentially/violent situations?		
Is counselling and support for the victims and witnesses of violence provided?		

12. Construction H&S Audits

In terms of the Construction Regulations, 2014, clients are required to conduct periodic audits and to provide the principal contractor with a construction H&S audit report within seven days of the completion of the audit. Contractors are required to conduct construction H&S audits at least every 30 days.

13. Features and Benefits of Construction H&S Audits

An effective audit process is an essential construction H&S management tool in any construction organisation. All well-structured and conscientiously conducted construction H&S audits provide an objective view of the actual status of construction H&S performance in the organisation itself and on the construction site, including identification of weaknesses, recognition of successes, evaluation of compliance, determination of the adequacy of the construction H&S policy and procedures against statutory requirements, world-class standards and social and ethical norms. A construction health and safety audit will, in general, look to assess the following factors:

- The strengths and weaknesses of the current construction H&S management system generally and its implementation on a construction project site.
- How the construction H&S management system performs within the aims of the construction organisation.
- Whether the construction organisation is fulfilling its legal obligations.
- Whether a proper construction H&S performance review system is in place.

A construction H&S compliance audit is one of the most important tools used on South African construction sites to ensure compliance with the Occupational Health and Safety Act of 1993, as amended, and the Construction Regulations, 2014, standards and by-laws. A construction H&S audit is a methodical, independent and documented assessment of the construction H&S management system and processes, measured against regulated criteria to make sure H&S standards are being upheld. Organisations should have a construction H&S management system in place to ensure H&S processes continue to be maintained. In addition to ensuring compliance, H&S audits also review the H&S documentation of the organisation to determine whether its record-keeping systems are adequate or whether they need to be more robust. Audits will also look beyond the current construction activity and evaluate the H&S training programme of the organisation.

A construction H&S audit is a proactive management approach to measure the H&S performance of the organisation and of its construction projects. This type of audit provides a basis for taking effective construction H&S planning decisions. It will help to:

- reduce the risk of personal traumas or injury to construction workers on construction sites
- prevent material loss from a construction business
- promote constructor worker morale
- ensure construction client confidence.

Broadly speaking, construction H&S audits serve two purposes, namely:

1. They are conducted to determine whether the construction organisation follows the applicable construction H&S legislative and regulatory framework.
2. They identify and highlight any weaknesses in the organisational and site H&S programmes and processes, providing guidance to interventions that improve construction H&S plans or corrective action that should be undertaken within specific timeframes.

14. Reasons for Conducting Construction H&S Audits

Reasons for conducting a construction health and safety audit include:

- To determine whether the construction H&S programmes and procedures in the organisation and on its construction sites are working.
- To verify that both construction workers and management are committed to and engaged in the construction H&S programmes.
- To verify that the construction processes follow the construction H&S policies, procedures and regulations of the organisation.
- To determine whether construction H&S programme activities are being documented properly.
- To uncover and identify potential hazards on the construction site.
- To evaluate the effectiveness of existing construction H&S management controls.
- To check routinely the construction H&S of the construction site and the equipment being used.
- To evaluate the adequacy of the construction H&S training and performance of supervisors and workers.

A health and safety audit is a more structured and thorough process than mere technical inspections or unplanned spot check inspections.

15. Difference between Construction H&S Audits and Construction Site Inspections

Table 22.1 sets out the key differences between construction H&S audits and site inspections.

Table 22.1: Difference between construction H&S audits and site inspections

Construction H&S audit	Construction site inspection
Aims to assess the construction H&S management system of the organisation	Aims to assess the use and effectiveness of hazard exposure control measures
A lengthy process that involves examination and scrutiny of the entire construction H&S management system	A short process that examines construction H&S practices on the construction site
Review of documentary evidence, backed up by observations and interviews of workers and personnel at all levels	Primarily based on observations involving limited paperwork and interviews of construction workers
Lengthy comprehensive report that records areas of concern and weakness in the construction H&S management system	Brief report identifying key corrective interventions that are needed
Detailed planning is required and requires considerable resources	Only limited planning and the main resource is the time of those doing the inspection
Typically done quarterly, every 6 months or annually	Usually done on a weekly basis
Aims to improve systems at a high level, with impact down to the operating level as a strategic tool to address long-term progress	Focuses on activities, tools and plant at and operational level, using remedial interventions that may address construction H&S management system faults

16. Types of Construction H&S Audit

Construction H&S audits are conducted in two ways, namely internally and externally. While internal audits are conducted by persons within the construction organisation, external auditors are appointed for external audits. As far as purpose, content, sequence and depth of investigation are concerned, both types of audit have similar lines of actions. The main differences lie in the approach. Internal auditors might have been trained as H&S auditors, yet they remain internal within the organisation. They may find it difficult to disclose the weakness of the management of the construction organisation and put it in writing in an auditor's report. The consequences could include a false sense of security and well-being within the organisation. Internal construction H&S auditing is practically meant as preparation for third-party external auditing. Third-party independent construction H&S audits are usually undertaken by expert agencies, comprising of teams of well-trained lead auditors, technical experts, legal specialists and others. Objectivity becomes the key motto and their disclosures are to be taken as bare facts and the basis on which further corrections and improvements are to follow.

The specific construction H&S audit identifies construction H&S hazards on a construction site. This type of audit is particularly useful in high hazard operations or where there is a high frequency of accidents. The specific audit is detail-oriented and time-consuming. To be effective, the construction H&S audit must be continuous and aligned with the day-to-day operations of the construction organisation.

An ongoing construction H&S audit process provides a platform from which management can obtain measurable and meaningful data about the construction H&S programmes of the organisation. On the other hand, a one-time, single construction H&S audit, such as an annual construction H&S audit, is ineffective in that it only provides a snapshot of the overall status of construction H&S programmes at a given point in time.

A construction H&S audit can involve a walk-through of the construction site, interviewing management or construction workers and reviewing the documentation of the organisation on the construction site.

17. The Construction H&S Audit Team

Given the importance of management commitment and involvement in the construction H&S effort, senior management must support and participate in the construction H&S audit process. They should endorse the process verbally and in writing to all construction workers in the organisation. In this way, construction workers become aware that the senior management of the organisation is serious about construction H&S audits and is committed to allocating appropriate resources to the effort.

There is no construction industry standard nor regulation that indicates who should conduct construction H&S audits. Variables such as the size of the construction business, the number and expertise of construction workers and special hazards relative to construction activities on site will dictate which construction workers are assigned to the audit programme. A large construction organisation may use a construction health and safety director or manager to implement and oversee the entire audit. An outside organisation may also be appointed to conduct the audit. Construction H&S consultants are commonly used.

18. Construction H&S Audit Checklists

The construction H&S audit must be documented in two major parts. The first part involves construction H&S checklists, while the second part involves the final construction H&S audit report. Checklists are an integral component of the overall construction H&S audit. These forms should suit the organisation and the type of construction H&S audit being undertaken. Key construction workers should be involved in the planning stages to ensure that all construction H&S programmes, operations and hazards are addressed. At a minimum, construction H&S audit checklists should be included for housekeeping; smoking; PPE; machinery/equipment and hand tools; fire health and safety; electrical health and safety; chemicals and hazardous substances; and emergency procedures.

The construction H&S audit checklist covers general construction H&S programmes and regulatory compliance; facilities and equipment; and specific hazards and operations. This checklist can help auditors identify construction H&S programmes required on the construction site. However, a checklist may not list all programmes or areas that need to be considered. Example 22.4 is an example of a construction health and safety audit checklist.

Example 22.4: Example of a construction H&S audit checklist

Risk Management Programmes	OK	Not OK	Tested	Regulatory Compliance Needed?	N/A
Construction Health and Safety Management and Regulatory Compliance					
Accident Investigations					
Accident and Injury/Illness Records					
Drug Testing/Substance Abuse					
Hazard Identification and Analysis					
Job Training (Specialised)					
Health and Safety Audits					
Health and Safety Meetings					
Health and Safety Programme					
Health and Safety File					
Safe Working Procedures (SWPs)					
Facilities and Equipment					
Building/Site Utilities (Electrical)					
Compressed Air Equipment					
Elevators, Stairs and Walking Surfaces					
Facilities and Equipment					
Emergency Response Plan					
Equipment Health and Safety (Policies, Maintenance, Repairs)					
Fire Prevention and Protection					
Housekeeping (Trash, Walkways, Aisles, Floor Maintenance, Tripping Hazards)					
Indoor Air Quality					
Life Health and Safety and Evacuation					

$\rightarrow$

Health and Safety for Visitors, Contractors etc				
Security				
Slip and Fall Prevention (Rugs/ Mats, Floor Care, Unmarked Elevation Changes etc)				
Stairs (Railings, Treads, Lighting)				
Vehicle Preventive Maintenance				
Specific Hazards or Operations				
Back Injury Prevention				
Confined Space Entry				
Construction Health and Safety				
Cranes (Chains, Hooks, Overhead Loads, Slings)				
Drivers (Selection, Training, Supervision)				
Electrical Safety				
Ergonomics Training				
Fall Prevention and Protection				
First Aid, CPR				
Forklift Safety and Operator Training				
Hazard Communication				
Hazardous Materials Inventory				
Hazardous Materials Training (Handling, Transportation, Storage, Disposal)				
Hearing Protection				
Industrial Hygiene Assessment and Monitoring				
LockOut/TagOut (LO/TO)				
Machine Guarding				
Material Handling				
Material Safety Data Sheets (MSDSs)				
Office Safety (Drawers, Electrical, File Cabinets Secured etc.)				
Painting Areas (Ventilation, Explosion-Proof Fixtures, PPE)				
Personal Protective Equipment				
Preconstruction Evaluation				
Scaffold and Ladder Safety; Work Platforms				
Shift Work/Work Schedules				
Thermal Conditions (heat, cold)				
Tool Safety				
Trenching				
Welding, Cutting, Hot Work				
Workplace Violence Protection Programme				

$\longrightarrow$

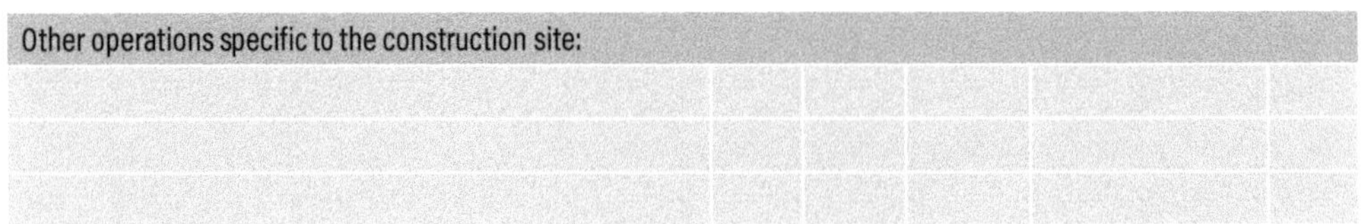

The second part of the documentation, the final construction H&S audit report, identifies the construction H&S audit findings; makes observations and recommendations; and offers an overall opinion. The report should provide detail on specific suggested enhancements to remedy deficiencies and should highlight serious and 'repeat' observations. The final construction H&S audit report should be communicated to management in a timely manner. Management should take ownership of the construction H&S audit results, and should approve improvements to construction H&S programmes, processes and equipment, both within the organisation and on its construction sites.

19. Review Questions

- What are the benefits of conducting a construction H&S inspection?
- How frequently should a construction H&S inspection be done?
- Why would one not include supervisors in the construction H&S inspection team?
- What should be in the final construction H&S inspection report?
- Why should an organisation undertake a construction H&S audit?
- What are the benefits of conducting a construction H&S audit?
- How frequently should a construction H&S audit be done?

20. Review Exercise

- Using the examples of construction H&S inspection and audit checklists, conduct a specific construction H&S inspection and an audit of the construction site, which requires the erection of external scaffolding to effect repairs to the roof of the building, as depicted in the following photograph.

Bibliography

Almond, P & Esbester, M. 2018. Regulatory inspection and the changing legitimacy of health and safety. *Regulation & Governance*, 12 (1): 46–63.

Canadian Centre for Occupational Health and Safety. nd. Effective workplace inspections, https://www.ccohs.ca/oshanswers/prevention/effectiv.html.

Goetsch, DL. 2013. *Construction safety and health*. Upper Saddle River: Pearson Education, Inc.

Griffith, A & Howarth, T. 2014. *Construction health and safety management*. Harlow: Pearson Education, Inc.

Haupt, TC. 2021. *Management of safety, health and environment in South Africa: A handbook*. Newcastle Upon Tyne: Cambridge Scholar Publishing.

Health and Safety Authority (Ireland). nd. Safety and health management, https://www.hsa.ie/eng/Topics/Managing_Health_and_Safety/Safety_and_Health_Management_Systems/.

Health and Safety Commission. 2007. *Managing health and safety in construction: Construction (Design and Management) Regulations 2007: Approved code of practice*. Sudbury: Health and Safety Executive Books.

Hinze, J. 2006. *Construction safety*. Gainesville: Jimmie Hinze.

Holt, ASJ. 2005. *Principles of construction safety*. Osney Mead, Oxford: Blackwell Science Ltd.

Lim, J, Kim, H & Park, Y. 2018. Review of the regulatory periodic inspection system from the viewpoint of defense-in-depth in nuclear safety. *Nuclear Engineering and Technology*, 50 (7): 997–1005.

Morantz, A. 2018. Construction site regulation and OSHA decentralization. *Members-only Library*.

National Association of Home Builders. 2007. *Home builders' safety program*. Washington: Builderbooks.

Walters, D, Quinlan, M, Johnstone, R & Wadsworth, E. 2016. Cooperation or resistance? Representing workers' health and safety in a hazardous industry. *Industrial Relations Journal*, 47 (4): 379–395.

CHAPTER 23

Multi-Stakeholder Construction Health and Safety

John J Smallwood

1. Introduction

Except for clients, construction health and safety agents (CHSAs), designers, principal contractors (PCs), contractors and the Occupational Health and Safety (OH&S) Inspectorate of the Department of Employment and Labour (DEL), the Construction Regulations are silent with respect to a range of stakeholders that influence construction health and safety (H&S).

However, section 10 of the Occupational Health and Safety Act 85 of 1993, as amended (OHSA) refers to articles, substances, manufacturers and suppliers, which effectively flags materials, plant and equipment. Further sections refer to the DEL's OH&S Inspectorate. Then, the Compensation for Occupational Injuries and Diseases Act 130 of 1993 (COIDA) refers to the Compensation Fund and mutual associations, which provide workers' compensation insurance cover. The National Building Regulations and Building Standards Act 103 of 1977, in turn, refers to municipal building inspection units.

The wide range of other stakeholders that influence construction H&S include, but are not limited to:

- Department of Public Works and Infrastructure (DPWI)
- Construction Industry Development Board (cidb)
- statutory councils
- National Home Builders Registration Council (NHBRC)
- professional, employer, and employee associations
- Construction Education and Training Authority (CETA)
- higher education institutions
- technical vocational education and training (TVET) colleges
- skills training institutions
- Quality Council for Trades & Occupations (QCTO)
- Green Building Council South Africa (GBCSA)
- National Treasury
- development agencies

- funding institutions
- media
- International Labour Organization (ILO)
- International Council for Research and Innovation in Building and Construction (CIB) Association of Researchers in Construction Safety, Health, and Well-Being (ARCOSH).

The range of stakeholders that influence construction H&S amplifies the need for a major environmental scan in terms of identifying the organisations that can and should influence construction H&S positively.

Furthermore, given the range of stakeholders involved in projects and their influence on project H&S, all projects should evolve integrated multi-stakeholder H&S plans as early as Stage 1, 'Project initiation and briefing'. Such H&S plans should indicate the required contributions of the primary stakeholders from stages 1 to 6 and the phase of the structure where their contribution will be most beneficial, and should include: clients; CHSAs; construction project managers (CPMs) or principal agents; designers; PCs; contractors; and the OH&S Inspectorate of the DEL.

2. Material Manufacturers and Suppliers

Materials are a major resource and one of 11 resources used in construction. Materials have varying characteristics and the materials process commences with inputs in the form of harvested or mined raw materials and the manufacturing process, and ends in the output material, which is then included in buildings, structures and infrastructure. The harvesting, mining, manufacturing and construction processes entail a range of H&S issues.

The following characteristics of materials impact on H&S as a result of any handling, storing, mixing, shaping and applying either as individual actions or as a combination of the shape, dimensions, surface area, sectional area, density (mass), centre of gravity, nature of edges, texture and constituents of materials. Therefore, materials manufacturers and suppliers can and do influence construction H&S.

Materials are the subject of legislation. Section 10 of the OHSA, 'General duties of manufacturers and others regarding articles and substances for use at work', is extensive in terms of requirements with respect to materials, including requirements impacting on:

- designers, manufacturers, importers, sellers or suppliers, who are required to ensure that, as far as is reasonably practicable, an article is safe and without risks to health when properly used

- erectors or installers of an article for use at work, who are required to ensure, as far as is reasonably practicable, that an article is safe or does not create a risk to health when properly used
- manufacturers, importers, sellers or suppliers of any substance for use at work, who are required to ensure as far as is reasonably practicable, that the substance is safe and without risks to health when properly used, and to provide the necessary information regarding the use of the substance at work, the H&S-related risks when used and the procedures to be followed should an accident involve the substance.

Various regulations in the Construction Regulations, 2014 refer to materials. Regulation 6, 'Duties of designers', (e) requires that designers refrain from including anything in the design of the structure necessitating the use of materials that are hazardous to the health and safety of persons, which can be avoided by substituting such materials. Although Regulation 9, 'Risk assessment for construction work', does not make specific reference to materials, it requires contractors to, among others:

(1)(a) conduct risk assessments prior to and during construction work to identify the hazards and risks to which persons may be exposed to

(1)(b) analyse and evaluate the hazards and risks

(1)(c) evolve a documented plan and safe working procedures to mitigate, reduce or control the hazards and risks

(2) analyse, evaluate and address ergonomic-related hazards in a risk assessment.

In theory, compliance with section 10 of the OHSA would nullify the need for Regulation 6(e) and, to a degree, (9)(1) and (2).

There is a paucity of, if any, South African information pertaining to beneficial H&S and ergonomics interventions relative to construction materials and components provided by material manufacturers and suppliers. However, developments over a period of three decades include the palletisation of stock or common bricks and off-loading by means of crane-mounted trucks in lieu of transport and delivery by flatbed or tip truck. The advent of prefabricated timber roof trusses has reduced the amount of work required at elevated heights relative to timber trusses assembled in situ. However, certain developments, such as pre-stressed hollow core concrete slabs and their contribution to productivity gains and cost reduction arising from their use in lieu of prestressed beam/rib and block or in situ reinforced concrete suspended slabs, have been documented. Although these benefits were highlighted, the substantial ergonomics and H&S benefits were not. These include elimination of manual handling, use of body force and other non-ergonomic actions as a result of the use of cranage and a lifting beam to hoist and position such slabs. The migration to water-based paints from solvent-based paints constitutes a further intervention to mitigate the existence of the hazardous chemical substances in construction materials.

3.   Plant and Equipment Manufacturers and Suppliers

H&S and ergonomics have been contributed to by the provision or availability of plant and equipment, and by design interventions relative to plant and equipment. Examples relative to H&S include mobile elevated working platforms (EWPs), which enable access to work at elevated heights and, in the case ergonomics, the design of earthmoving equipment to enhance operator body posture and movements, and to mitigate the impact of vibration on the body.

Issues such as whole-body vibration (WBV), which concerns the transmission of vibrations into the human body and hand-arm vibration (HAV), are considered such serious occupational health hazards that the European Community issued an H&S directive dealing with them. A range of actions can be taken to reduce the risk of WBV and HAV, including, among others, selecting the right equipment, isolating the vibration, changing work processes, operator good practice, and education and training. Selecting the right equipment and isolating the vibration are two interventions that plant and equipment manufacturers have contributed to in terms of the design of plant and equipment.

4.   OH&S Inspectorate of the DEL

OHSA mandates the OH&S Inspectorate of the DEL to enforce the provisions of the Act and the regulations promulgated thereunder.

For example, inter alia:

- Section 24 requires the reporting of certain types of workplace-related incidents to the inspectorate. However, this does not include motor vehicle accidents (MVAs) in the course of employment, which must be reported to the South African Police Service. This fragmentation of reporting results in the understating of construction fatalities in the South African construction industry by the DEL, as MVAs during employment contribute approximately 40 per cent of all fatalities.
- Section 25 requires medical practitioners to report diseases arising from patients' employment to the inspectorate. However, a prerequisite is that medical practitioners are familiar with occupational diseases and the likely origin thereof.
- Section 29 authorises inspectors to conduct unannounced inspections, question persons, request and examine documentation, seize documentation and inspect various aspects of the workplace.
- Section 30, inspectors can issue prohibition, contravention and improvement notices, as deemed appropriate.

- Section 31 authorises inspectors to investigate any incident to determine whether it is necessary to hold a formal investigation.
- Section 32, an inspector may subpoena any person to appear before a public inquiry and a written report is required upon conclusion thereof, which is then submitted to the National Prosecuting Authority for a decision about whether to prosecute or not. Although several construction Section 32 inquiries have been conducted relative to high profile incidents, reports have not been released to the public. An exception is the section 32 investigation report into the Injaka bridge collapse on 6 July 1998. Given that many accidents experienced in South African construction tend to be replicated, and that invaluable lessons could be learnt from section 32 investigation reports, it would be helpful if section 32 inquiries are concluded and the related reports be published.

Application for a permit to do construction work in terms of Construction Regulations Regulation 3 is discussed in an earlier chapter. Similarly, notification of construction work in terms of Construction Regulations Regulation 4 is covered in an earlier chapter.

The OH&S Inspectorate of the DEL faces several challenges, including, among others:

- a plethora of workplaces
- a capped number of inspectors
- downstream intervention, which is reactive, as opposed to proactive
- limited, if any, influence relative to H&S culture, in terms of values, vision, mission, goals, purpose and assumptions.

Importantly, *H&S cannot be inspected into construction.* Deliberate interventions must be introduced by those executing construction activities.

5. Municipal Building Inspection Units

The main function of municipal building inspection units is to ensure that all construction within their jurisdiction complies with the National Building Regulations and Building Standards Act and that construction conforms to the conditions of the approved building plans or documents. The following are the areas of responsibility of such inspection units:

- conducting inspections during and after construction relative to the following:
  - setting out of the structure
  - foundations before constructing the foundations themselves
  - open drainage
  - the roof
  - completion of the building or structure

- issuing of hoarding permits, excavation permits, blasting permits, overhead crane permits and demolition permits
- enforcement in terms of taking action with respect to deviations and non-compliance in terms of the approved plans or the National Building Regulations and Building Standards Act
- issuing of occupancy certificates
- providing expert advice to the public
- attending to building-related complaints.

The responsibilities clearly indicate the linkages to H&S and the role of municipal building inspection units in construction H&S. Furthermore, it is notable that there were no approved building plans and a non-approved excavation plan application for the Tongaat Mall project, which experienced a collapse on 19 November 2013 that resulted in two fatalities and twenty-nine injured (Broughton, 2013).

6. Workers' Compensation Insurance Providers

Similarly to other countries, South Africa also has its own workers compensation legal framework.

6.1 Compensation Fund

The Compensation Fund (CF) is overseen by the DEL and the CF's operations are confined to the insurance of employers against their liabilities in terms of COIDA. The CF pays compensation to permanent and casual workers, trainees and apprentices who are injured or contract a disease in the course of their work, and lose income as a result.

Compensation benefits will not be paid if:
- an accident is reported to the employer more than 12 months after the accident or death, or after the disease was diagnosed
- the injured person is off work for three or less days, in which case the CF will only pay medical expenses
- the accident resulted from the negligence or wrong doing of the injured person, unless the injured person is seriously disabled as a result of or dies in the accident, in which case the CF will pay compensation
- the injured unreasonably refuses or wilfully neglects to have medical treatment.

6.2 The Federated Employers Mutual Assurance Company (RF) (Pty) Ltd

The Federated Employers Mutual Assurance Company (RF) (Pty) Ltd (FEM) was established as a mutual insurer in 1936 and, upon the introduction of the Workmen's Compensation Act 30 of 1941, FEM was granted a license to continue to transact worker's compensation insurance for the construction industry. Although FEM's business operations are confined to the insurance of employers against their liabilities in terms of section 30, 'Mutual associations' of the COIDA, FEM undertakes a range of interventions related thereto. FEM funds the employment of H&S advisors in the employ of the master builder associations MBAs and the South African Forum of Civil Engineering Contractors (SAFCEC). Other interventions include:

- the provision of statistics relative to the employers they provide insurance to
- sponsoring of bursaries, industry H&S competitions, star grading programmes, H&S awards, conferences, seminars, workshops and facilitating of H&S-related research.

Section 85 of the COIDA provides information with respect to tariffs (rates) and rebates:

(1) if an employer's business is designed, equipped, organised or conducted in a manner that is calculated to prevent accidents, and the number of accidents and the related expenditure are, or are likely to be, less than the norm, the Director-General may assess that employer at a lower tariff of assessment than the norm

(2) if the accident record of an employer during a particular period is less favourable than the norm and it is likely that the status quo will probably continue, the Director-General may assess such an employer at a higher tariff of assessment than the norm

(3) if the accident record of an employer during a particular period is more favourable than the norm, the Director-General may grant such an employer a rebate on any assessment paid or payable by him.

The loading or rebate is applied two years retrospectively because both CF and FEM use a two-year retrospective period to allow a reasonable time for all the claims costs to be reflected. This period was determined by claims history and experience as being the most reasonable period. FEM calculates the rebate, or loading, based on a single year's actual premium and claims costs incurred. The claims cost expressed as a percentage of the premiums paid constitutes the loss ratio. A favourable loss ratio, which is between 0 per cent and 62 per cent, could result in a merit rebate (subject to conditions). A loss ratio between 63 per cent and 64 per cent has no effect on rebates or loadings. A loss ratio of 65 per cent and higher attracts a loading. Therefore, FEM can and does influence construction H&S performance by granting rebates and the application of penalties, as can the CF.

7. Department of Public Works and Infrastructure

The Department of Public Works and Infrastructure (DPWI) is well positioned to promote H&S by:

- including H&S as a project value
- funding H&S on projects undertaken as a public sector client
- appointment of CHSAs regardless of whether a permit to do construction work is required or not
- prequalifying PCs in terms of H&S and requiring the appointment of candidate construction H&S practitioners on DPWI projects.

Furthermore, DPWI is the custodian of programmes such as the Contractor Incubator Programme, administered under the Emerging Contractor Development Programme, and the Expanded Public Works Programme, which constitutes an opportunity to include H&S training in various forms, such as H&S induction, hazard identification and risk assessment (HIRA) and personal protective equipment (PPE) courses, and toolbox talks, therein.

8. Construction Industry Development Board

In terms of the Construction Industry Development Board Act 38 of 2000, the Construction Industry Development Board (cidb) established a national register of contractors, which categorises contractors in a manner that facilitates public sector procurement and promotes contractor development. It is notable that H&S is not included in the criteria for registration, which constitutes a major opportunity for the cibd to engender focus on H&S as part of their development mandate.

The cidb produced an industry status report in 2004 that referred to H&S and the landmark *Construction Health & Safety Status & Recommendations* H&S status report. The first *The cidb Construction Industry Indicators: Summary Results* report was published in 2007 and the last in 2015. H&S was one of five groups of indicators included in the reports and, thus, provided an annual update with respect to H&S according to a range of stakeholders.

The cidb developed the *Standard for Health & Safety Plans and Auditing Requirements (Grades 2 to 9)*, which addresses the requirements for H&S specifications and plans; guidelines for preparing, and the contents of, H&S plans; a checklist to evaluate the adequacy of an H&S plan, and a checklist to audit the implementation of an H&S plan.

Lastly, the cidb developed the draft *Standard for Primary Health Assessments for Construction Works Contracts (Grades 5 to 9)*, which provides guidance and includes a proforma 'Primary Health Assessment for Construction Workers'.

9. Statutory Councils

In South Africa, most if not all the built environment professional disciplines are regulated by various discipline-specific statutory councils.

9.1 Council of the Built Environment

The Council of the Built Environment (CBE) is the umbrella statutory council for the built environment.

The CBE's contributions to H&S include:

- stakeholder forums, which have included H&S breakaway sessions
- transformation indaba, which have addressed H&S
- an H&S think tank, which involved a range of stakeholders
- undertaking of H&S-related research projects.

The Health, Safety, Public Protection and Universal Access (HSPPUA) is one of five transformation collaborative committees. The goals and objectives of HSPPUA relative to H&S are:

- set timeframes to resolve specific H&S issues
- advocate for sector compliance with H&S legislation
- include H&S in stages 1 to 3
- explore avenues to ensure that National Treasury and the DPWI include H&S from project inception stages with clients and implementing agents.

The CBE has the mandate to ensure that all statutory built environment councils include H&S in their scope of services of registration categories from stages 1 to 6. Currently, the South African Council for the Project and Construction Management Professions (SACPCMP) is the only council that does and, then, in a comprehensive manner.

9.2 South African Council for the Project and Construction Management Professions

CBE mandated SACPCMP, in terms of the Project and Construction Management Professions Act 48 of 2000, to register construction H&S professionals following a report of the cidb, which highlighted the need for professional registration of construction H&S practitioners because of, inter alia, the finding that there was a lack of competencies and no formal registration process. This, in turn, led to the identification of three categories of registration, namely Professional Construction Health and Safety Agent (Pr CHSA), Construction Health and Safety Manager (CHSM) and Construction Health and Safety Officer (CHSO). Registration rules were then gazetted for these three categories for commencement on 1 June 2013, in the case of Pr CHSA, and 1 August 2013, in the case of both CHSM and CHSO. The Construction Regulations, 2014, in turn, make provision for the appointment of CHSAs and require the appointment of either part-time or full-time CHSOs. The impact of these CHS categories on improving overall construction H&S of the industry in South Africa is debatable. Arguably, the only category that should be registered is that of the Pr CHSA, which, in the European context, was intended to be developed as a new built environment professional. The other categories can be accommodated under the provisions of OSHA for H&S representatives, provided they became full-time and suitably qualified and trained CHS employees.

The CHSA Scope of Services, in turn, states that CHSAs, CHSMs and CHSOs are expected to be experienced and knowledgeable relative to the following:

- construction project H&S management systems
- construction H&S management
- construction H&S performance measurement and monitoring, and construction H&S continual improvement.

10. National Home Builders Registration Council

The National Home Builders Registration Council (NHBRC) is a regulatory body of the home building industry established in 1998 in accordance with the provisions of the Housing Consumers Protection Measures Act 95 of 1998. NHBRC's mandate is to protect the interests of housing consumers and to ensure that builders comply with the prescribed building industry standards as contained in the *Home Building Manual*.

The Housing Consumers Protection Measures Act states that the NHBRC's role is to, among others, establish and to promote ethical and technical standards in the home building industry and to improve structural quality in the interests of housing consumers and the home building industry. Given that H&S is an ethical issue, the NHBRC should:

- include H&S as a value of the council
- include H&S criteria for registration
- produce a Health and Safety Guide for Home Builders
- include the review of H&S on site during inspections
- promote and/or deliver construction H&S training
- institute an H&S competition for NHBRC registered home builders.

Furthermore, the non-achievement of structural quality has implications for H&S during construction and the occupancy of homes.

11. Professional Associations

Several professional associations and learned societies have been established.

11.1 The Association of Construction Health and Safety Management

The Association of Construction Health and Safety Management (ACHASM) is a registered non-profit company and a recognised SACPCMP voluntary association established to provide all those working in the construction H&S field with an advisory and representative body. ACHASM is committed to promoting the professional interest of construction H&S practitioners within the built environment in terms of the Construction Regulations.

ACHASM fulfils the following functions with respect to CHS:

- promotes the services and skills of members to a range of stakeholders
- provides assistance with requirements for training, knowledge, experience and CPD
- provides best practice information and technical assistance
- provides opportunities for involvement in the development of procedures and standards for accreditation of courses, which affect member requirements or the practice of their discipline
- facilitates H&S-related research especially relative to the performance of H&S practitioners
- arranges CPD events, such as monthly presentations, seminars and an annual two-day summit
- offers discounted rates for members; negotiates a specialised cost-effective group professional indemnity insurance scheme for all professionally registered ACHASM members on an annual basis
- arranges free annual subscription to the African OS&H magazine
- promotes ethical practices and holds members to account, where such standards are not met.

Furthermore, membership contributes five points towards the SACPCMP application to register scorecard.

11.2 Chartered Institute of Building

The Chartered Institute of Building (CIOB) is the world's largest and most influential professional body for construction management and leadership. The CIOB has 15 hubs globally, one being sub-Saharan Africa. The CIOB representative is based in Cape Town.

In terms of tertiary education, the CIOB's Undergraduate Education Framework schedules the learning outcomes for programmes from Level 4 up to honours degree programmes at Level 6. One of six themes, section 2.3 'Health, Safety and Wellbeing', in turn, addresses four themes in the form of legislation and practice; personal responsibility; management; and wellbeing and safety culture relative to the three levels.

The CIOB undertakes campaigns that include H&S-related campaigns such as, for example, 'Change in our Sites' campaign in 2003, launched to identify how the situation needed to change in the construction industry to the raise standard of site conditions, and the 'Mental Health in Construction' campaign in November 2020.

The CIOB has initiated several H&S-related research studies. Reports include *The State of Well-being in the Construction Industry* published in December 2017 and *Understanding Mental Health in the Built Environment* published in May 2020. The discussion paper *Reimagining construction: The vision for digital transformation, and a roadmap for how to get there*, jointly prepared by Autodesk and CIOB, and published in 2019, clearly articulated the role on Industry 4.0 technologies in enhancing, among others, H&S.

CIOB H&S-related guidelines include *Coronavirus Resources and Guidance*, published in 2021.

11.3 South African Institute of Occupational Safety and Health

The South African Institute of Occupational Safety and Health (Saiosh) is recognised by the South African Qualifications Authority (SAQA) as the professional body for the registration of occupational H&S professionals in South Africa in terms of the National Qualifications Framework Act 67 of 2008. Saiosh is also a recognised SACPCMP voluntary association. Although Saiosh is a generic professional association, many of its members are involved in the construction industry.

Saiosh fulfils the following functions:

- promotes the services and skills of members to a range of stakeholders
- provides assistance with requirements for training, knowledge, experience and CPD
- provides free e-learning and legal updates
- arranges CPD events such as conferences, seminars, webinars and workshops
- negotiates a specialised cost-effective group professional indemnity insurance scheme for members
- arranges free annual subscription to the *SHEQ Management* magazine
- acts as a lobby group for occupational H&S legislation and standards by interacting and liaising with the DEL and other government bodies on behalf of its members.

Furthermore, membership contributes five points towards the SACPCMP application to register scorecard.

12. Green Building Council South Africa

The Green Building Council South Africa (GBCSA) developed the Green Star Socio-Economic Category pilot in 2014 in the form of a technical manual in an endeavour to move beyond the traditional green focus to include broader sustainability impacts, the application of which is encouraged in order to assess and improve the socioeconomic attributes of projects.

One point is awarded where a primary health programme for construction site employees has been implemented, which includes:

- conducting primary health assessments for all consenting construction-related employees, including subcontractor employees, at least once during the construction phase and preferably at the beginning of the project, and referring employees for further medical examination or treatment when problems are identified
- conducting a basic health awareness programme for construction-related employees, particularly addressing the key health issues identified in the primary health assessments, for example HIV and AIDS; tuberculosis; drug and alcohol abuse; and malaria, depending on the project
- a minimum expenditure of R500 for both the assessment and awareness components per employee involved in construction
- where construction staff accommodation or labour camps are provided for the project, such staff accommodation must be designed, constructed and operated according to best practice standards.

However, prerequisites include that the project complies with the OHSA, the Construction Regulations and other related regulations promulgated in terms of section 43 of the OHSA, and that design HIRAs are conducted to enable the amendment of the aspects of design that are hazardous.

13. Employer Associations

There are two associations in South Africa that cater to the needs of employers in the construction industry.

13.1 Master Builders South Africa

Master Builders South Africa (MBSA), the building construction umbrella employer association, has a national H&S coordinator. The various master builders associations (MBAs), in turn, employ H&S advisors funded FEM. The MBAs visit projects to raise awareness with respect to H&S, provide H&S advice and conduct H&S inspections. Their H&S committees include all the relevant stakeholders and meet to discuss, among others, H&S competitions, H&S injury statistics, new H&S legislation and regulations and topical issues. They also organise regional H&S competitions, contribute to the national H&S competition and manage the national H&S star grading programme, participation in which contractors regard highly. Further interventions include the arranging and hosting of H&S conferences, seminars and workshops.

13.2 South African Forum of Civil Engineering Contractors

The South African Forum of Civil Engineering Contractors (SAFCEC) is an established and recognised employer association in the civil engineering sector of the construction industry. A dedicated safety, health, environment, risk and quality (SHERQ) department, staffed with well qualified and experienced construction health, safety and environment practitioners, offers the following services driven by member mandate and requirements:

- best practice development
- conflict resolution skills
- stakeholder liaison
- incident investigations
- training
- site visits and site audits
- H&S campaigns
- ad hoc SHERQ support.

The SHERQ department is funded by FEM and maintains a cordial relationship with that entity. Furthermore, SAFCEC collaborates with other industry stakeholders in terms of various H&S initiatives to enhance H&S performance in the civil engineering sector.

14. Employee Associations (Unions)

South Africa adopts a tripartite approach to the development of H&S legislation and regulations, which includes government, employers and unions. Therefore, unions can influence H&S legislation and regulations.

Unions have a major role to play in terms of representing workers in terms of:
- bridging the communication gap and assisting in terms of understanding
- providing advice and legal assistance, if necessary

- participating in H&S inspections, accident investigations and accident inquiries
- supporting workers in the case of victimisation.

A further role is that of providing H&S information and H&S education and training, and raising and maintaining H&S awareness.

Campaigning for an improvement in H&S and partnering with employer associations or employers should be holistic, and, therefore, include aspects of H&S such as primary health promotion and well-being, which, in turn, highlight the importance of unions in addressing issues, such as the number of working days per week and the number of working hours per day. When negotiating generic agreements with employer associations or employers, H&S should be addressed. Generic agreements and H&S agreements should address, for example, the following:

- provision of protective wear, such as overalls and PPE
- provision of facilities, such as change rooms, eating areas, showers and toilets
- provision of primary health promotion
- the right to investigate accidents, conduct H&S inspections and participate in H&S inquiries
- provision of incentives for H&S
- H&S recognition.

15. Development Agencies

Development agencies such as the Coega Development Corporation in the Eastern Cape Province of South Africa, which also doubles as an implementing agency, are in a unique position to influence and contribute to construction H&S because they invariably have in-house H&S departments. Influence and contributions should:

- commence with the development of an integrated multi-stakeholder project H&S plan
- include a comprehensive H&S component in the client brief
- identify an appropriate project duration
- include an infusion of H&S into the design process
- identify an appropriate construction procurement system
- include contract documentation that refers to H&S
- provide detailed H&S preliminaries
- include prequalification of PCs in terms of H&S
- reference H&S during the pre-tender site meeting
- ensure that the PC has made adequate financial allowances for H&S.

16. Funding Institutions

Funding institutions have a major role to play in construction H&S because of their potential influence in terms of scheduling criteria relative to H&S as a prerequisite for developers receiving the funding.

The environmental, health, and safety (EHS) guidelines of the International Finance Corporation (IFC), which is part of the World Bank Group, are technical reference documents with general and industry-specific examples of good international industry practice. When one or more members of the World Bank Group are involved in a project, these EHS guidelines are applied as required by their respective policies and standards. The general EHS guidelines are designed to be used together with the relevant industry sector EHS guidelines, which provide guidance to users on EHS issues in specific industry sectors. For complex projects, the use of multiple industry-sector guidelines may be necessary.

The National Treasury Standard for Infrastructure Procurement and Delivery Management refers to H&S in this 66-page document as follows:

- Stage 3: 'Preparation and briefing or prefeasibility' states that the strategic brief shall as necessary: a) confirm the scope of the package and identify any constraints, including those relating to occupational H&S.
- Stage 4: 'Concept and viability or feasibility' states that the concept report shall as necessary: f) include a baseline risk assessment (BRA) assessment for the package and an H&S plan, which is required in terms of the requirements of the Construction Regulations, and g) contain a risk report linked to the need for further surveys, tests, other investigations, consents and approvals, if any, during subsequent stages and identified health, safety and environmental risk.
- Contract Management, where the person responsible for administering the contract shall as necessary report on a monthly basis on the following: b) the number of improvement, contravention and prohibition notices issued by the CHSAs, and c) incidents reportable in terms of the Construction Regulations, briefly indicating the nature of the incident.

The National Treasury's Framework for Infrastructure Delivery and Procurement Management refers to H&S in the 32-page document as follows:

- Stage 2: 'Concept' (ii) The Concept Report should as a minimum, provide the following information: f) Include a baseline risk assessment for the project, and an H&S plan, which is a requirement of the Construction Regulations.
- '5.3 Operations and Maintenance Processes: Operations Management Plan (OMP)': (ii) The plan must include: f) Risks and Occupational Health and Safety (OHS) provisions.

The National Treasury's standard and framework both require review, especially given the opportunity to infuse H&S into their ten project stages (stage number in parentheses): project initiation (0); infrastructure planning (1); strategic resourcing (2); prefeasibility (3a); preparation and briefing (3b); feasibility (4a); concept and viability (4b); design development (5); design documentation (6): production information (6a), and manufacture, fabrication and construction information (6b); works (7); handover (8); and package completion (9).

The Focus on: National Treasury Standard for Infrastructure Procurement and Delivery Management refers to H&S eight times in the 88-page document as follows:

- Figure 2 features the 'H&S agent' as part of the 'delivery team', and states that the H&S agent assumes the statutory responsibilities imposed by the Construction Regulations and leads H&S risk management compliance processes. Figure 2 also features the client and refers to clients having to comply with legislation, including occupational H&S and environmental legislation.

- It is notable that 'H&S services' are included as one category of services provided by the delivery team in a discussion relative to Figure 2.

- Further discussion refers to the client, designer and the contractor as being responsible for ensuring compliance with the provisions of the Construction Regulations, assigning of the client's functional responsibilities to H&S agents and that H&S professionals must be registered in terms of the Project and Construction Management Professions Act.

- Budget, schedule, quality and performance characteristics required from the completed works and rate of delivery are recorded as tangible objectives among the primary objectives relating to the delivery and maintenance of infrastructure. Buildability, in turn, is referred to as an intangible objective. However, 'improving H&S performance' is recorded as secondary objective, along with, among others, work opportunities for small, medium and micro enterprises, alleviation of poverty, local economic development and development of cidb-registered contractors.

The recognition of H&S as a secondary objective is notable in a negative sense, as a result of H&S being a recognised project parameter at least equal to the other project parameters in terms of importance, namely cost, developmental objectives, environment, productivity, quality and time.

17. Higher Education Institutions

17.1 Universities

The cidb advocates that construction H&S should be offered as a separate subject or as an identifiable component in a subject within the civil engineering, construction management, construction project management and quantity surveying programmes offered at universities. Furthermore, architectural programmes should include H&S as an identifiable component of design and professional practice, and the evaluation of design projects, working drawings and details should include H&S as one of the criteria. Ideally, designing for construction H&S and design HIRA should be included in such programmes because failure to do so would mean that graduates would be unaware of H&S when in practice, which is not ideal.

A study titled the *Tertiary Built Environment Construction H&S Education Framework Study* supported the inclusion of several critical construction H&S aspects in the construction H&S modules of built environment programmes.

Significantly and of concern is that, except for construction management programmes, most tertiary level built environment education programmes only address construction H&S to a limited extent.

17.2 Technical vocational education and training colleges

The courses offered at technical vocational education and training (TVET) colleges are vocational or occupational in nature. This means that the student receives education and training with a view to participation in a specific range of jobs, employment or entrepreneurial possibilities.

The H&S courses presented by TVET colleges should emulate the scope of the construction H&S modules of built environment programmes at tertiary level, depending on the discipline concerned. Given that constraints generally exist in terms of credit values, construction H&S should be referred to in all the modules of construction-related courses offered by TVET colleges, if a dedicated construction H&S module cannot be accommodated.

18. Construction Education and Training Authority

As the construction sector education and training authority, CETA should be playing a major role in construction H&S by:

- providing leadership in terms of the inclusion of H&S in construction skills training, construction-related courses offered by TVET colleges and tertiary level built environment education
- funding of H&S learnerships

- sponsoring of bursaries relative to diploma, advanced diploma, undergraduate, honours, coursework masters construction H&S studies, construction H&S oriented masters by research and doctoral studies
- funding of construction H&S experiential learning
- funding industry construction H&S research.

19. Quality Council for Trades and Occupations

QCTO is a quality council that was established in 2010 in terms of the Skills Development Act 97 of 1998, as amended in 2008. QCTO is South Africa's public entity responsible for quality assurance and the oversight of the design, accreditation, implementation, assessment and certification of occupational qualifications, part-qualifications and skills programmes. QCTO also offers guidance to skills development providers, private and public, and assessment centres, who must be accredited by QCTO in order to offer occupational qualifications.

The National Certificate: Construction Plant Operations, NQF Level 02 and the SAQA Qualification ID 65789, states that qualifying learners will be capable of:
- applying H&S principles in plant operations
- understanding plant operations in the construction environment
- performing plant operation-related functions
- adhering to plant operation personnel procedures
- understanding mechanical and leverage principles to operate earthmoving, transport and/or ancillary plant.

It is notable that learning outcomes 1 and 4 make direct reference to H&S and that learning outcomes 2, 3 and 5 relate to H&S performance.

20. Skills Training

The NQF Level 3 'Construction Health and Safety' (SAQA ID 77063) learnership is intended to develop a broad understanding and knowledge of H&S, and of environmental legislation and controls. The focus of the programme is on enabling learners to responsibly, independently and effectively manage themselves on construction sites with regard to generic H&S issues. Learners should be able to identify and evaluate occupational safety, hygiene and environmental factors in occupational environments that may have a detrimental effect on the health and safety of people in such environments.

21. Media

Construction print media editors agree that the media can positively influence H&S.

The advent of social media presents a further opportunity in terms of H&S. Social media is an important communication medium for professional associations, contributes to H&S endeavours and the advancement of H&S, and raises the profile of the professional association concerned.

22. International Labour Organization

While promoting individual and collective rights at work, social protection and occupational H&S, the ILO encourages social dialogue and supports an open and constructive industrial relations policy between governments, employers and workers. The ILO conducts research relative to these and a range of other issues in the ever-changing world of work, and publishes the results in the form of timely and authoritative publications, reports, training manuals, CD-ROMs, videos and e-books. ILO research contributes to enhancing awareness of crucial labour and employment issues in subject areas such as H&S. While focusing primarily on employment in the global economy, the ILO publications programme provides relevant research findings and practical solutions to workplace problems for workers and employers in developing, transitioning and industrial economies.

Notable construction-related publications include, among others, *Safety and Health in Construction* in January 1992; Safety, *Health and Welfare on Construction Sites: A training manual* in January 1995; *Good Practices and Challenges in Promoting Decent Work in Construction and Infrastructure Projects* in October 2015 and *Code of Practice on Safety and Health in Construction* in February 2022.

23. Research Organisations

23.1 International Council for Research and Innovation in Building and Construction

CIB was established in 1953 as an association whose objectives were to stimulate and facilitate international cooperation and information exchange between governmental research institutes in the building and construction sector, with an emphasis on those institutes engaged in technical fields of research.

CIB has since developed into a worldwide network of over 5 000 experts from approximately 500 member organisations active in the research community, in industry or in education, who cooperate and exchange information in over 50 CIB commissions covering all fields in building and construction related research and innovation.

CIB members are institutes, firms and other types of organisations involved in research, or in the transfer or application of research results, and appoints experts to participate in CIB commissions. Individuals can also be a member of and participate in a commission. CIB commissions initiate collaborative research projects, organise conferences, congresses, meetings, symposia and produce publications. Publications include proceedings emanating from conferences, congresses, and symposia, reports and, more recently, research roadmaps.

The CIB Working Commission W099, 'Safety Health and Wellbeing in Construction', is committed to the advancement of safety, health and well-being (SHW) of construction workers. The tools necessary to accomplish this include designing, preplanning, training, management commitment and the development of an H&S culture. Several CIB W099-endorsed conferences and W099 conferences have been staged in South Africa and successfully promoted H&S-related research among and collaboration between academics, industry practitioners, postgraduate students and research institutions. Furthermore, the conferences were infused with international contributions in the form of keynote addresses, scientific papers and workshop presentations. Overall, the conferences contributed to human resource development, knowledge transfer, research outputs and accrual of CPD points for members of professional associations and registered persons.

23.2 Association of Researchers in Construction Safety, Health, and Well-Being

Over a period of time, it has become apparent that there are several researchers conducting construction SHW-related research on the African continent evidenced by, among others, papers being published and personal approaches to the founder members of ARCOSH to collaborate. Furthermore, a common challenge on the African continent is the lack of construction-specific H&S legislation and regulations, as well as statistics, and a lack of collaboration in terms of construction SHW research.

ARCOSH was launched on 11 October 2018, its mission being to promote research and development relative to construction SHW, and to contribute to the achievement of optimum construction SHW on projects by:
- promoting research and development to improve construction legislation, systems, processes, procedures, and practices
- forming partnerships with research organisations, built environment statutory councils, professional associations, registration boards, tertiary education institutions and other built environment stakeholders
- promoting collaboration and coordination of construction SHW research between countries.

The inaugural ARCOSH conference was held in Cape Town on 3–4 June 2019. The first day included keynote addresses and scientific papers, while the second day, the 'Industry Day', addressed topical subjects courtesy of academics, and industry practitioners to a mixed audience, which included industry practitioners.

24. Review Questions

- Is the current level of construction H&S education, training and CPD optimum?
- What are the benefits of a multi-stakeholder approach to construction H&S?
- Should the Construction Regulations be expanded to include construction stakeholders other than those currently included? A pragmatic response is required.

25. Review Exercises

- Compile an integrated multi-stakeholder project H&S plan that indicates the primary project stakeholders' potential contributions, based on the contents of this and previous chapters.
- Compile a 750-word article titled *The multi-stakeholder nature of construction H&S*.

Bibliography

The Association of Construction Health and Safety Management (ACHASM). 2021. *Association of Construction Health and Safety Management (ACHASM)*. Port Elizabeth: ACHASM.

Association of Researchers in Construction Safety, Health, and Well-Being (ARCOSH). 2018a. *Introduction*. Port Elizabeth: ARCOSH.

Association of Researchers in Construction Safety, Health, and Well-Being (ARCOSH). 2018b. *Vision statement*. Port Elizabeth: ARCOSH.

Broughton, T. 2013. Mall tragedy of broken rules. *Natal Mercury* 21 November 2013: 1.

Building Inspection Unit. 2022. *Building development management information booklet*. Johannesburg: Johannesburg Building Inspection Unit, https://www.joburg.org.za/Campaigns/Documents/2016%20documents/Building%20Development%20Management%20-%20BuildingInspectUnit.pdf.

Chartered Institute of Building (CIOB). 2018. *CIOB undergraduate education framework*. Bracknell: CIOB.

Chartered Institute of Building (CIOB). 2022. *Who we are*. Bracknell: CIOB, https://www.ciob.org/about/who-we-are.

Civilution. 2016. *Focus on: National Treasury standard for infrastructure procurement and delivery management.* Midrand: South African Institution of Civil Engineering (SAICE).

Construction Industry Development Board (cidb). 2004. *SA construction industry status report - 2004.* Pretoria: cidb.

Construction Industry Development Board (cidb). 2009. *Construction health & safety status & recommendations.* Pretoria: cidb.

Construction Industry Development Board (cidb). 2017. *Draft standard for primary health assessments for construction works contracts (grades 5 to 9).* Pretoria: cidb.

Construction Industry Development Board (cidb). 2018. *Draft standard for health & safety plans and auditing requirements.* Pretoria: cidb.

Department of Labour. 2002. *Section 32 investigation report into the Injaka bridge collapse of 6 July 1998.* Pretoria: Department of Labour.

Edwards, D & Holt, GD. 2005. *A guide to: Whole-body vibration.* Loughborough: The Off-highway Plant and Equipment Research Centre.

International Finance Corporation (IFC). 2007. *Environmental, health, and Safety (EHS) general guidelines.* Washington: IFC.

National Treasury. 2015. *Standard for infrastructure procurement and delivery management.* 1st edition. Pretoria: National Treasury.

National Treasury. 2019. *Framework for infrastructure delivery and procurement management.* 2nd edition. Pretoria: National Treasury.

Smallwood, JJ. 2006. The Practice of Construction Management. *Acta Structilia* 13 (2): 62-89.

Smallwood, JJ. 2010. Construction management health and safety (H&S) course content: Towards the optimum. In: P Barrett, D Amaratunga, R Haigh, K Keraminiyage & C Pathirage (eds) *Proceedings of the CIB World Building Congress*, Salford Quays (14–17 May 2010), http://www.disaster-resilience.net/cib2010/files/papers/908.pdf.

Smallwood, JJ. 2015. The need for the inclusion of construction health and safety (H&S) in architectural education to assure healthier and safer construction. In: *Proceedings of CIB W099 Workers and Society through Inherently Safe(r) Construction*, Belfast, Northern Ireland, United Kingdom (10-11 September 2015): 169-179.

Smallwood, JJ. 2018. The impact of posts on interest in a professional construction health and safety management association Facebook page. In: *Proceedings of the Joint CIB W099 and TG59 International Safety, Health, and People in Construction Conference*, Salvador, Brazil (1–3 August 2018): 206-212.

Smallwood, JJ. 2019. Tertiary built environment construction health and safety education in South Africa. In: P Manu, F Emuze, T Saurin & B Hadikusumo (eds) *Construction health and safety in developing countries.* London, UK: Routledge: 152-66.

Smallwood, JJ & Bester, D. 2020. Proliferation of health and safety documentation in construction. In: *Proceedings The 8th International Conference on Construction Engineering and Project Management (ICCEPM 2020).* Hong Kong SAR, China (7–8 December 2020): 243–248.

Smallwood, JJ & Venter, D. 2001. The role of the media in South African Construction Health and Safety (H&S). *The Australian Journal of Construction Economics and Building,* 1 (1): 43–52.

South African Council for the Project and Construction Management Professions (SACPCMP). 2013a. *Application for registration.* Midrand: SACPCMP.

South African Council for the Project and Construction Management Professions (SACPCMP). 2013b. *Registration rules for construction health and safety agents in terms of section 18 (1) (c) of the Project and Construction Management Professions Act, 2000 (Act No. 48 of 2000).* Midrand: SACPCMP.

South African Council for the Project and Construction Management Professions (SACPCMP). 2013c. *Registration rules for construction health and safety managers in terms of section 18 (1) (c) of the Project and Construction Management Professions Act, 2000 (Act No. 48 of 2000).* Midrand: SACPCMP.

South African Council for the Project and Construction Management Professions (SACPCMP). 2013d. *Registration rules for construction health and safety officers in terms of section 18 (1) (c) of the Project and Construction Management Professions Act, 2000 (Act No. 48 of 2000).* Midrand: SACPCMP.

South African Institute of Occupational Safety and Health (Saiosh). 2022. *Membership,* https://www.saiosh.co.za/page/About_membership.

South African Qualifications Authority (SAQA). 2022. *National Certificate: Construction Plant Operations,* https://allqs.saqa.org.za/showQualification.php?id=65789.

CHAPTER 24

Scheduling for Construction Health and Safety

Theo C Haupt

1. Introduction

Construction planning and scheduling refers to the process through which a construction contractor maps out, as far as reasonably possible, what will take place during a construction project, what resources are needed, when key construction activities will take place, which construction personnel or workers will be involved and anything else pertinent to the construction project. A construction project schedule is a high level document that comprises the planned dates for performing scheduled construction activities and the planned dates for meeting scheduled milestones on the construction project. While not a requirement of the Construction Regulations, 2014, scheduling for construction health and safety (H&S) done at the same time as construction project scheduling will enhance overall construction H&S management performance. Comprehensive planning, organising, controlling, monitoring and reviewing H&S responsibilities are all essential for the success of any construction project and the key to achieving healthy and safe working conditions on the construction site. The adage of 'plan, schedule and manage' resonates well with the notion of scheduling for construction H&S.

2. Integration of Construction H&S with the Project Schedule

Research has shown that one of the leading causes of construction project failure is poor project planning, suggesting that by not having a plan, project team members are usually unaware of what is expected on the project until it is too late. The same is true for construction H&S. Therefore, to improve construction project H&S planning, it is necessary to integrate H&S management into project scheduling. Hazardous situations arise at various times during project execution and the schedule or programme should consider these exposures as part of construction H&S planning and management on the project. However, research and experience has shown that current H&S planning approaches do not take updates of the construction project schedule, made to address the dynamic changes that occur on construction sites as a result of expected hazard exposures and H&S controls, into account. Therefore, to create more effective construction site-specific H&S plans, it is important to integrate these plans into the construction project schedules. In this way, high risk periods in the project schedule from a construction H&S point of view will enable more effective mitigation of potential hazard exposures in advance. The use of an integrated project schedule enables, for example, the early prediction of exposure to fall hazards and the areas on the construction site where this exposure is likely to occur. Risks to exposure vary across the project timeline and, therefore, it makes good sense to use the findings of the construction hazard identification and risk assessment (HIRA) to map the significant hazards onto the project schedule.

Construction H&S should not be treated as a separate specialist aspect of a project; rather it should be treated as part of the construction process itself. When one accepts this view, construction H&S becomes the responsibility of everyone involved in the construction project and on the site.

3. Integrated H&S Project Schedule as a Communication Tool

It has been suggested that poor communication can have a negative impact on successful construction project execution. Studies have found that about 18 per cent of all construction projects fail because of a lack of communication. Efficient communication between all the parties involved in construction project delivery is, therefore, a critical project success factor. Integrating the findings of the HIRA, during which the significant project hazards and risks have been identified, assessed and prioritised into the overall construction project schedule, provides an opportunity to timeously communicate with clients, consultants, contractors, subcontractors, suppliers and manufacturers about construction H&S as the project is executed. In this way, all parties involved in the project are able to prepare in advance for any potential exposures that might occur. Similarly, the schedule also serves as a powerful tool to communicate information about potential exposures to the workers before they execute hazardous construction activities.

4. Content of an Integrated H&S Project Schedule

In order to integrate H&S into the construction project schedule when it is being prepared, it will be necessary to consider whether there are any construction activities that will affect the health and safety of others working at the construction site at any given time, especially those identified in the HIRA. Consider, for example, the following:

- Access to the various workplaces on the construction site in terms of which construction activities will need to occur where and when.

- Organise the construction activities to ensure that everyone working on the site who needs to use a scaffold or other means of access has time to do so and then plan to ensure that access to the scaffolding will be safe and suitable for the specific use.
- Where a specialist contractor or subcontractor is going to work on the project, their requirements will need to be determined, especially whether these requirements might affect the health and safety of themselves and others working on the site, and that work needs to be scheduled well in advance.

5. Example of an Integrated H&S Project Schedule

Example 24.1 is an extract of an integrated H&S construction schedule that shows how, prior to each construction activity in the schedule, construction H&S items have been included, showing when each item should be executed as part of the project planning.

Example 24.1: Extract from a construction project schedule for repairing a collapsed boundary wall into which construction H&S items have been integrated in order to demonstrate that it is possible and, indeed, advisable to schedule for construction H&S on all projects

INTEGRATED CONSTRUCTION HEALTH AND SAFETY SCHEDULE: REPAIRING COLLAPSED BOUNDARY WALL. SEZELA, KZN					
ITEM ID	DESCRIPTION	DURATION	START	FINISH	COMMENT
1	PRE-PROJECT CONSTRUCTION HEALTH AND SAFETY MEETING	1 DAY	23 Mar	23 Mar	Ensure that all pre-project construction health and safety procedures and requirements have been met. such as prequalification of all parties, appointments have been made in writing, letters of good standing (LOGS) etc.
2	PROJECT CONSTRUCTION HEALTH AND SAFETY KICK-OFF MEETING	1 DAY	30 Mar	30 Mar	Review construction H&S plan, required construction H&S policy, procedures and requirements that include construction H&S project induction, security arrangements, construction H&S method statements, risk assessments, safe working procedures, certificates of fitness, notifications, process for compiling construction H&S file etc.
3	Site establishment	1 day	07 Apr	07 Apr	

ITEM ID	DESCRIPTION	DURATION	START	FINISH	COMMENT
4	Site meeting with client's principal agent	0.5 days	08 Apr	08 Apr	Include discussion about client's construction H&S goals for project and procedures, including content and procedures for compiling the construction H&S file.
5	CONSTRUCTION H&S PROJECT INDUCTION (ongoing)	0.13 DAYS (about 1 hour)	08 Apr	08 Apr	All construction workers on the construction site will have to attend the construction H&S project induction when they first come onto the site, whether starting work there or being transferred from other projects. Similarly, all visitors and project participants must attend the construction H&S project induction when they arrive on site. Proof of induction must be provided to all attendees and a register must be kept as part of the project records, NOT part of the construction H&S file.
6	Site clearance	2 days	08 Apr	10 Apr	
7	SITE CLEARANCE TOOLBOX TALK	0.13 DAYS	08 Apr	08 Apr	Explain safe working procedures (SWPs), proper use of tools, equipment and personal protective equipment (PPE).
8	Removing vegetation, hedges, shrubs and bush	1 day	08 Apr	08 Apr	
9	Taking out trees and removing and grubbing up	1 day	12 Apr	12 Apr	
10	Demolitions	3 days	13 Apr	15 Apr	
11	DEMOLITIONS TOOLBOX TALK	0.13 DAYS	13 Apr	13 Apr	Explain demolitions method statement, SWPs, emergency procedures, proper use of tools, equipment and PPE.
12	Breaking up and removing brick paving	1 day	13 Apr	13 Apr	
13	Breaking and removing half brick walls	1 day	14 Apr	14 Apr	
14	Breaking and removing reinforced concrete	1 day	15 Apr	15 Apr	
15	Excavations	4 days	18 Apr	21 Apr	

$\rightarrow$

ITEM ID	DESCRIPTION	DURATION	START	FINISH	COMMENT
16	EXCAVATIONS TOOLBOX TALK	0.13 DAYS	18 Apr	18 Apr	Explain excavations and backfilling method statement, excavation and confined spaces SWPs, emergency procedures, proper use of tools, equipment and PPE.
17	Stripping average 150 mm thick layer of topsoil	1 day	18 Apr	18 Apr	
18	Open face excavations in earth	1 day	19 Apr	19 Apr	
19	Setting out foundations and erect profiles	0.5 days	20 Apr	20 Apr	
20	Excavation for trenches	1.5 days	20 Apr	21 Apr	
21	Excavations for manholes	1 day	21 Apr	21 Apr	
22	Soil poisoning to bottoms and sides of excavations	0.5 days	21 Apr	21 Apr	

6. Review Questions

- Why should construction H&S be integrated into the project schedule?
- How can the integrated construction H&S safety project schedule improve communication on the project?
- How is construction project management improved by integrating H&S with the project schedule?

7. Review Exercise

Use the extract of an integrated health and safety construction schedule in Example 24.1 to complete the schedule by including the construction activities after the excavations have been completed. This schedule should include the site handover at the completion of the project and the construction H&S file handover.

Bibliography

Goetsch, DL. 2013. *Construction safety and health*. Upper Saddle River: Pearson Education, Inc.

Griffith, A & Howarth, T. 2014. *Construction health and safety management*. Harlow: Pearson Education, Inc.

Health and Safety Commission. 2007. *Managing health and safety in construction: Construction (Design and Management) Regulations 2007: Approved code of practice.* Sudbury: Health and Safety Executive Books.

Hinze, J. 2006. *Construction safety.* Gainesville: Jimmie Hinze.

Holt, ASJ. 2005. *Principles of construction safety.* Osney Mead, Oxford: Blackwell Science Ltd.

National Association of Home Builders. 2007. *Home builders' safety program.* Washington: Builderbooks.

CHAPTER 25

Construction Health and Safety Supply Chain Management

Theo C Haupt

1. Introduction

To achieve exceptional and effective construction health and safety (H&S) performance on construction projects and sites requires management of the entire construction supply chain. The Construction Regulations, 2014 repeatedly refer to the appointment of participants in the construction project delivery process and the requirement for their appointment to be based on either being both competent and adequately resourced or being competent to deliver the service required on the construction project. In many cases, the requirement to only be competent is specifically stated. While not specifically referring to the process of prequalification in the Construction Regulations, 2014, it is implied that each participant will be subjected to being prequalified by the party for whom they will be delivering a particular service with respect to construction H&S, in terms of either both competence and resources or only competence, prior to being approved and appointed to the construction project.

2. Parties to be both Competent and Adequately Resourced

The Construction Regulations, 2014 specifically require that the following parties in the construction project delivery process be both competent and adequately resourced, namely:

- principal contractor
- client-appointed health and safety agent
- other contractors.

3. Parties to be Competent

The Construction Regulations, 2014 require the following parties to be competent, namely the:

- construction manager
- construction H&S officer
- person
  - conducting and compiling the risk assessment
  - conducting any required training
  - conducting periodic inspections
  - preparing the fall protection plan
  - responsible for temporary works, including those involved in erection, supervision and inspection
  - supervising excavations
  - supervising and controlling demolitions
  - supervising the use of explosives during demolitions
  - responsible for scaffolding
  - responsible for the operational compliance plan
  - responsible for erection, operation and inspection of suspended platforms
  - supervising rope access
  - inspecting of material hoists
  - supervising bulk mixing plant
  - designing and erecting cranes
  - operating construction vehicles and mobile plant
  - supervising stacking and storage
  - inspecting fire equipment
  - serving on the construction H&S technical committee.

4. Assessment of Competence and Resources

Table 25.1 shows how the competencies of client-appointed H&S agents, construction entities, contractors, suppliers, manufacturers and designers can be assessed, as advocated by the Health and Safety Executive in the United Kingdom and adapted for South African construction purposes. The responsibility for ensuring competence and/or adequate resources is a non-delegable one, except in the case of a client who appoints a construction H&S agent.

Table 25.1: How competencies can be assessed

Aspect	Criteria	Standard to be achieved	Examples of the evidence that would demonstrate meeting the required standard
1	H&S policy and management of H&S in the organisation	Have and implement an appropriate policy that is reviewed regularly and signed off by senior or top management. The policy must be relevant to the type and scale of the kind of construction work that is done. It must set out clearly the responsibilities for H&S management at all levels in the entity who are prequalified.	A signed and current copy of the H&S policy that must show the dates of revisions and who the responsible person is. Guidance on how to write a construction H&S policy is provided in Chapter 3 of this book.
2	Arrangements	This aspect should outline the arrangements for construction H&S management within the entity and must be relevant to the type and scale of the construction work that is undertaken. The arrangements should set out how the duties and responsibilities required by the Occupational Health and Safety Act, as amended, and the Construction Regulations, 2014 will be discharged.	A concise but clear explanation of how the entity will manage construction H&S and implement the provisions of the H&S policy. It must show how the duties and responsibilities will be discharged. For example, processes for recordkeeping, such as medical certificates of fitness. Evidence of process for keeping letters of good standing from the Compensation Commissioner current. The arrangements should include a construction workforce communication plan and include the elements outlined in chapters 17, 18, 19 and 23 of this book.
3	Competent advice, both corporate- and construction-related	Everyone in the entity must have easy access to competent construction H&S advice from within the entity itself. The party providing the advice should be able to provide general guidance on construction H&S matters.	Details of the party providing construction H&S advice, whether internally (preferred) or externally. Evidence of the type of advice given in the preceding year or so and how that advice was implemented.
4	Training and information	Training arrangements should be made and implemented to ensure that everyone has the necessary construction H&S knowledge, skills and understanding to be able to discharge their duties and responsibilities on construction projects and sites. There should be construction H&S refresher training included in the programme, which could include continuing professional development (CPD) training and/or lifelong learning designed to keep everyone updated on new developments, changes to legislation and regulations and improvements to construction H&S practices and procedures. Senior or top management must also be included in the training programme.	Evidence in the form of construction H&S training records that demonstrate a construction H&S culture and should include records of the types of training, topics covered, certificates of attendance or attendance registers, induction programme, samples of toolbox talks and details of a CPD programme.

Aspect	Criteria	Standard to be achieved	Examples of the evidence that would demonstrate meeting the required standard
5	Individual qualifications and experience	Everyone in the entity is expected to have appropriate qualifications and experience for their spheres of construction H&S duties, responsibilities and activities, especially those who supervise the construction activities of others.	Evidence of qualifications and experience of relevant persons, such as construction and site managers, construction H&S advisors and supervisors. Evidence of registration with relevant bodies, such as the South African Council for the Project and Construction Management Professions (SACPCMP). Other key roles in the form of an organogram should be populated with details of relevant qualifications and experience. Details of the proportion of the construction workforce who have construction H&S qualifications and experience, especially those who are H&S representatives or have construction H&S responsibilities on site. Details of qualifications, experience and professional body registration and membership and/or certification in construction professional disciplines. The design team must show the proportion of their staff who also have relevant construction H&S training. Similarly, client-appointed agents must have appropriate construction H&S qualifications, experience and training, as well as professional registration with SACPCMP. Clear commitment to construction H&S training and CPD records of everyone.
6	Monitoring, audit and review	There should be a system for monitoring construction H&S policies, procedures and arrangements, auditing them at specific intervals and an ongoing review.	Evidence of formal audits, discussions and reports to senior or top managers, recent monitoring and management response and copies of construction H&S site inspection reports, medical certificates of fitness and letters of good standing from the Compensation Commissioner.
7	Construction workforce engagement and involvement	An established system of consulting with the entire workforce on construction H&S matters should be implemented.	Evidence to show how consultation is done, including records and minutes of construction H&S committee meetings, records of involvement of construction H&S representatives in inspections and audits, including their identities.

Aspect	Criteria	Standard to be achieved	Examples of the evidence that would demonstrate meeting the required standard
8	Accident and incident reporting and enforcement procedures, and follow up investigations	All prescribed and mandatory records for at least three years. There are types of documentary records that must be kept for up to 40 years, such as exposure to asbestos and lead during construction operations. There should be a system for reviewing all incidents, as well as documenting the action that was taken as a consequence. There should also be records of all enforcement actions that were taken against the construction entity over the past five years, the remedial action that was taken and the interventions that were subsequently developed, introduced and implemented.	Evidence showing the recordkeeping and investigation procedures with respect to accidents and incidents. Records of at least two accidents/incidents and the actions taken to prevent their recurrence. Records of any enforcement action taken over the past five years and the resultant remedial action taken. Any statistics of the incidence rates of major injuries, reportable incidents and cases of ill-health and significantly hazardous occurrences during the previous three years that the entity has in place and uses could also be presented as evidence. Records of incidents and accidents involving labour-only subcontractors and independent contractors should also be produced.
9	Subcontracting/ consulting/ manufacturing/ supplier/ service provider procedures	There should be arrangements in place for appointing construction H&S competent and resourced subcontractors and/or consultants, manufacturers, suppliers and other service providers. Similarly, there should be procedures in place to ensure that subcontractors also have the same kinds of arrangements in place for making similar appointments. There should be a system of monitoring the construction H&S performance of subcontractors, consultants, manufacturers, suppliers and other service providers.	Evidence of the procedures, with examples, to ensure that subcontractors, consultants, manufacturers, suppliers and other service providers are construction H&S competent and adequately resourced.
10	Hazard elimination and risk control	Arrangements should be made for identifying significant hazards on the construction project and construction site to ensure proper mitigation of exposure to potential hazards.	Evidence in the form of examples where hazards were eliminated following the National Institute for Occupational Safety and Health (NIOSH) hierarchy of control and how any remaining risks are controlled. Summary of how risk exposures were reduced on a construction project and/or site.
11	Risk assessment	There should be procedures in place for carrying out risk assessments and for developing and implementing interventions to reduce exposure to ensure a safe and healthy construction site.	Evidence that the hazard identification and risk assessment (HIRA) process is followed in the entity and how risks are be controlled. Samples of risk assessments/safe working procedures/ method statements.

Aspect	Criteria	Standard to be achieved	Examples of the evidence that would demonstrate meeting the required standard
12	Cooperating with others and coordination with other contractors	There should be procedures in place to demonstrate how cooperation and coordination of the construction work is achieved in practice on the construction site, and how the construction workers are engaged and involved in developing H&S method statements and safe working procedures.	Sample risk assessments, procedural arrangements, minutes and notes of project team meetings, and indications of how the entity coordinates with other contractors/entities.
13	Welfare provision	Demonstration of what appropriate welfare facilities will be put in place before the commencement of any work on the construction site and how they will be accessed.	Examples of a construction H&S policy commitment, contracts with the suppliers of portable toilet and ablution facilities and details of the type of facilities provided on previous projects.
14	Client-appointed construction H&S agents	Demonstration of cooperation, coordination and communication to achieve the duties and responsibilities set out in the Construction Regulations, 2014. Registration with the SACPCMP.	Actual examples, rather than generic procedures. Proof of registration with the SACPCMP.
15	Work experience	Detailed evidence of relevant experience in the kind of construction being undertaken and the structure being erected.	A record of recent projects/contracts should be kept with the contact details of clients and others who can verify that work was executed with due consideration for construction H&S. Letters of completion from clients. Where there are any shortcomings in experience, an explanation should be given about how these shortcomings will be overcome on the project.

5. Developing the Competence of Construction Workers

Table 25.2 provides an example of how the competence of general construction workers can be developed over a period of 24 months.

Table 25.2: Development of general construction worker competencies over 24 months

Stage of development	Construction H&S knowledge
Starting work on the construction site	■ Can identify hazards on the construction site. Has knowledge of basic construction H&S legislation and regulations. ■ Knows the security arrangements and procedures on the site. ■ Knows the emergency arrangements and evacuation procedures.
Trainee	■ Can identify deteriorating conditions on the construction site that could lead to increased exposure to hazards. ■ Increasing knowledge of significant hazards and risks, and the approved methods for controlling the exposure to risks.
Trained	■ Can participate fully and confidently in H&S consultations on the construction site.
Qualified	■ Has a working knowledge of the statutory requirements with respect to PPE, reporting and the duties and responsibilities of construction workers on the site.
Competent	■ Possesses sufficient knowledge to instruct other trainee workers on H&S systems of executing construction activities on the site.

6. Review Questions

- Who should be responsible for ensuring that everyone who is appointed is either competent or adequately resourced, or both, to be able to execute their function on the construction project without threatening the health and safety of other parties involved in the construction delivery process?
- Why is this process of prequalification important and necessary?
- When selecting a supplier, what would be the most important H&S considerations to be taken into account?
- What is meant by a non-delegable responsibility. with reference to construction H&S appointments?

7. Review Exercise

Conduct the prequalification exercise for a subcontractor who must execute an electrical installation in a 100-room hotel that is being refurbished.

Bibliography

Health and Safety Commission. 2007. *Managing health and safety in construction: Construction (Design and Management) Regulations 2007: Approved code of practice.* Sudbury: Health and Safety Executive Books.

National Association of Home Builders. 2007. *Home builders' safety program.* Washington: Builderbooks.

CHAPTER 26

Construction Health and Safety Closeout Report

John J Smallwood

1. Introduction

One of the issues with respect to construction health and safety (H&S) is the recurrence of incidents. Therefore, the need for project H&S closeout reports is amplified to ensure that relevant information, such as lessons learnt, is captured, documented and distributed among all the project stakeholders.

Legislation in the form of The Scope of Work for Categories of Registration of the Project and Construction Management Professions, 2019 states that construction managers should prepare and present a contract closeout report; construction project managers (CPMs) should prepare and present the project closeout report, and construction health and safety agents (CHSAs) should prepare the consolidated construction project H&S closeout report.

The requirements of the Construction Regulations, 2014, among others, the explicit duties of clients, CHSAs, designers and contractors, and the implied duties of CPMs, cost engineers or quantity surveyors (QSs), are such that it would be beneficial if they are all included in the closeout report.

A further motivation for a closeout report is the striving for continuous improvement within the context of total quality management and the realisation of better practice.

2. Custodian of and Rationale for the Process

The question is who the custodian of the process should be. The individual project stakeholders could each evolve their own closeout reports because of differing objectives and, in theory, each stakeholder should review their own performance. However, from a client perspective, clients influence project H&S because of the baseline risk assessment (BRA), provision of the respective versions of the H&S specification, review of the designers' reports, inputs provided to the design process, and procurement- and construction-related interventions, such as auditing the H&S performance of the contractors. Furthermore, clients fund projects and may have project H&S objectives and goals, and, therefore, there is a direct incentive to optimise and improve H&S performance. CHSAs fulfil the function of the client's agent in terms of H&S and are reliant on future appointment, and, therefore, overall project H&S performance and improvement is important. The challenges faced by construction project managers (CPMs) are multiple in the sense that:

- they are responsible for at least all six project stages, including integrating the design, procurement and construction processes
- clients may have project H&S objectives and goals
- decisions during the early project stages, namely stages 1 to 3 and procurement, affect H&S performance during construction
- H&S performance affects performance relative to the other project parameters
- a closeout report is recorded as a deliverable during Stage 6 in the construction project management identity of work.

The focus of designers should be on designing for H&S, design hazard identification and risk assessment (HIRA) being a key related intervention to mitigate the experience of design-originated hazards during construction. This requires, among others, the development of a construction H&S hazard library. The focus of cost engineers or QSs should be on facilitating financial provision for H&S, determining the cost of H&S and the development of a construction H&S hazard library. Contractors need to improve H&S compliance and performance, which affect performance relative to the other project parameters, reputation, image and market competitiveness. Furthermore, they need to mitigate the experience of hazards during construction, which requires, among others, the development of a construction H&S hazard library; the determination of the cost of H&S; and the production of a closeout report.

3. Contents of a Closeout Report

A project H&S closeout report should follow the format of most reports:

- cover page
- dedication
- acknowledgements
- origin of the report
- scope of the report
- executive summary
- table of contents
- introduction
- overview of the project
- project scope
- H&S key performance indicators

- project performance assessment and analysis
- challenges and risks faced, and resolution thereof
- lessons learned
- recommendations for future projects
- conclusions.

3.1 Closeout report introduction

The introduction to the closeout report should indicate the:
- project name
- project location
- project commencement and finish dates
- project value
- number of storeys
- basement details, if applicable
- height and footprint of the building
- floor area per storey
- key project stakeholders in the form of the client, CHSA, CPM, respective designers, QS, principal contractor (PC)
- number of the various types of subcontractors including:
  - direct contractors
  - nominated subcontractors
  - peak and mean number of workers
- key plant such as:
  - the number of tower cranes
  - personnel and material hoists
  - mast climbing work platform
  - peak number of mobile cranes
  - concrete pumps
  - and batching plant in terms of capacity.

3.2 Overview of the project

The project overview should provide information with respect to the following:
- demolition, if it took place
- earthworks
- type of structure and structural frame
- enclosing fabric
- roof structure and covering
- lifts, escalators and travelators
- type of internal compartmentation, such as masonry walls and drywalling

- major types of finishes
- site works, such as roads, parking, landscaping and swimming pools.

3.3 Scope of the project

The scope of the project should provide information with respect to the:
- purpose of the project
- project objectives, including H&S
- key client requirements, including H&S
- project participation in terms of small, medium and micro enterprises and communities adjoining the project site
- other socioeconomic objectives
- prequalification criteria, if any
- contract award criteria including preferential procurement points
- number of units, if a housing complex, block of apartments, residences or hotel
- phased handovers if any
- key milestones.

3.4 H&S performance measures

H&S performance measures include leading and trailing indicators. Leading indicators, which are predictors of H&S performance, should be identified by the PC during Stage 4, 'Tender documentation and procurement', while undertaking project planning and budgeting for H&S. However, the client and/or the CHSA, in conjunction with the CPM or principal agent, may develop such indicators in the form of client requirements and include them in the contractor H&S specification at Stage 4. The various elements of an H&S programme provide information with respect to the possible leading indicators and categories thereof such as, for example:
- education and training that includes management HIRA training, supervisor HIRA training, worker HIRA training and H&S representative training
- management commitment and involvement that includes attendance of H&S committee meetings, and inclusion of H&S on the agenda of site management and subcontractor meetings.

Trailing indicators, which are reactive measures that measure negative outcomes, should be reviewed monthly during a project. The fatality rate per 100 000 fulltime equivalent workers is a globally understood measure. The disabling injury incidence rate is the number of disabling injuries per 100 workers based on a mean of 2 000 hours worked per year, that is, 40 hours per week for 50 weeks. A disabling injury, in turn, is an injury that results in the injured worker losing a shift or more after the day on which the injury occurred. The medical aid injury incidence rate is the number of injuries that require medical aid treatment per 100 workers based on a mean of 2 000 hours worked per year, for example, stitches or sutures to close a wound by a medical practitioner. The first aid injury incidence rate is the number of injuries that require first aid per 100 workers based on a mean of 2 000 hours worked per year, for example, removal of a speck of dust from an eye by a first-aider. The severity rate is the number or working days lost per 100 workers based on a mean of 2 000 hours worked per year. Hours worked without a lost-time injury is a globally understood measure. The workers' compensation claims ratio represents the percentage that claims constitute the total amount of workers' compensation insurance paid. This ratio cannot be readily computed and provided, and certainly not monthly because of the time lag between the occurrence of certain classes of injury and the finalisation of the costs of treatment, rehabilitation and other forms of compensation relative thereto.

In terms of the assessment of the H&S performance measures, the trailing (outcome) performance measures can be compared to the contractor's lowest, average and highest trailing performance, and the industry average trailing performance or, if the contractor is part of a benchmarking group, the group's lowest, average and highest trailing performance.

The leading performance measures can be deliberated relative to the trailing (outcome) performance measures, while remembering that the former were, in many cases, relative to the contractor as an organisation, and not the management, supervisors and workers that worked on the project being assessed.

3.5 Assessment of criteria per stage

Ideally, the performance of the respective stakeholders per criteria can be assessed quantitively or qualitatively. The former would require a questionnaire survey, while the latter would require an interview protocol. However, a focus group discussion involving the key project stakeholders would be more practical and would enable discussion, debate and possible consensus.

In terms of the six stages advocated by the South African Council for the Project and Construction Management Professions, the following may be useful:

3.5.1. Stage 1: Project initiation and briefing

The H&S competency of the CPM or principal agent and CHSA is critical since they should provide H&S leadership at this stage, In addition, Stage 1 is the ideal stage to influence construction H&S during the downstream stages, that is, what was notable, and what was lacking.

The following questions should be addressed:

- Was the H&S brief as an aspect of the client brief informative and effective?
- To what extent did it inform and influence H&S during the downstream stages?
- Were the client's H&S objectives and H&S goals appropriate, and to what extent were they achieved?
- Were the client's H&S requirements adequate and what was the extent of compliance therewith?
- Was the client's BRA adequate, and were hazards and risks not identified?
- Was the designer H&S specification adequate, informative and detailed, and were hazards and risks encountered that were not identified?
- Was the multi-stakeholder project H&S plan comprehensive, and did the respective stakeholders contribute to the H&S endeavours and the H&S performance on the project?

3.5.2. Stage 2: Concept and feasibility

Was the indicative project programme realistic, in general, and was it appropriate in terms of managing H&S on site?

From a pricing for H&S perspective, was the budget for H&S on the part of the cost engineer/QS adequate relative to the PC's estimate? Were additional H&S preliminaries items added due to variation orders having an impact on H&S?

From a concept design perspective did the designers influence H&S during the downstream stages, by undertaking, among others, design HIRA and, thus, reflect H&S competency? From a concept design and feasibility perspective, did the cost engineer/QS influence H&S during the downstream stages and thus reflect H&S competency?

3.5.3. Stage 3: Design development

During Stage 3, designers should conduct comprehensive design HIRAs and evolve a detailed designer report indicating how the hazards and risks were addressed, and what were the residual risks. The adequacy of the design HIRA process will be determined by the encountering of design-originated hazards and risks during construction, if any, which required an amendment to the design, or additional contractor H&S-related interventions.

3.5.4. Stage 4: Tender documentation and procurement

The contractor H&S specification should provide information to the PC of the client requirements and the residual hazards, and provide other relevant H&S information. The adequacy thereof can be assessed post Stage 5, 'Construction documentation and management', and Stage 6, 'Project closeout', by documenting additional client H&S requirements, hazards and risks identified, and H&S information requested, which was not included therein.

The submission of H&S documentation to local authorities, such as the 'notification to commence with construction work' and the 'permission to encroach on streets and/or sidewalks', can be assessed in terms of whether the required range of H&S documentation was submitted or not, and whether it was approved on the first submission or not.

The construction work permit application can be assessed in terms of its completeness and whether it was approved on first submission, or not.

The assessment of facilitation of financial provision for H&S can be undertaken in terms of the approach, that is detailed H&S preliminaries, lump sum or provisional sum, and whether the approach was satisfactory in terms of appropriate allowance and mitigation of additional claims.

The review of the assessment of the PC's financial provision for H&S should focus on the effectiveness of the assessment process relative to the facilitation of financial provision for H&S. Was the client able to determine whether the PC made adequate financial provision for H&S, or not?

The review of the assessment of the contractor's H&S competencies and resources should focus on the contractors' performance during stages 5 and 6 relative to the former process.

The assessment of the site layout can be undertaken in terms of the adequacy and appropriateness of the site fencing or hoarding, entrances, guardhouses, site offices, materials' stores, welfare facilities, living accommodation, plant, stockpiles, temporary roads, stockpiles, on-site manufacturing and site layout.

3.5.5. Stage 5: Construction documentation and management

The review of the assessment, discussion, negotiation and approval of the PC's H&S plan should be undertaken in terms of the duration of the process, the possible amendment thereof and its effectiveness, that is, was the H&S plan appropriate in terms of the actual construction process and its activities, or not? Was the first version approved?

The assessment of the H&S component of the site handover brief can be undertaken in terms of the encountering of unexpected H&S issues.

Temporary works design should be assessed in terms of the process followed and the success thereof in the form of, among others, no failures. Temporary works execution can be assessed in terms of conformance to design, productivity relative to the process, conformance of the permanent structure in terms of quality standards, schedule performance and no failures.

Welfare facilities should be assessed in terms of compliance with the Construction Regulations, Facilities Regulations, H&S plan adequacy and maintenance.

The assessment of any revised H&S risk profile, specification and budget, when scope of work changes, can be assessed in terms of a lack of incidents and accidents.

The implementation of the H&S plan can be assessed courtesy of the monthly site audits; however, such audits reflect the site conditions and H&S practices on site at a particular point in time. Accidents, injuries and incidents constitute a further indicator, although the absence of an incident or accident does not indicate that a site is healthy and safe.

The effectiveness of design HIRA can be determined by reviewing the encountering of any design-originated hazards during Stage 5.

3.5.6. Stage 6: Project closeout

The development and outcome in terms of the H&S file can be assessed in terms of the process followed through the project stages, the contributors, the content and the original framework included in the designer H&S specification constituting the benchmark.

The assessment of the implementation of the H&S plan is as per that for Stage 5.

Aspects of H&S related to operations and maintenance should be assessed in terms of the availability of related manuals and the addressing of H&S therein.

Table 26.1 presents a summary of the H&S performance measures and the form of assessment that a project H&S closeout report should address in terms of leading and trailing indicators as well as the form of assessment at each of the six stages.

Table 26.1: Summary of contents of a construction H&S closeout report

Aspect	Form of assessment
H&S performance measures	
Leading	The trailing (outcome) performance measures relative to the leading performance measures
Trailing	Fatality rate per 100 000
	Disabling injury incidence rate
	Medical aid injury incidence rate
	First aid injury incidence rate
	Severity rate
	Hours worked without a lost time injury
	Claims ratio
Criterion/Stage	**Form of assessment**
Stage 1: Project initiation and briefing	
CPM/Principal Agent's H&S competency	Adequacy
CHSA's H&S competency	Adequacy
H&S brief	Adequacy
Client's H&S objectives	■ Appropriateness ■ Extent of achievement

$\rightarrow$

Criterion/Stage	Form of assessment
Client's H&S goals	<ul><li>Appropriateness</li><li>Extent of achievement</li></ul>
Client's H&S requirements	<ul><li>Adequacy</li><li>Extent of compliance</li></ul>
Baseline risk assessment	<ul><li>Adequacy</li><li>Hazards and risks not identified</li></ul>
Designer H&S specification	<ul><li>Adequacy</li><li>Hazards and risks not identified</li></ul>
Multi-stakeholder project H&S plan	Adequacy
Stage 2: Concept and feasibility	
Indicative project programme	Appropriateness in terms of H&S
Budget for H&S	Adequacy
Designer's H&S competency	Adequacy
Cost engineer/QS's H&S competency	Adequacy
Stage 3: Design development	
Design HIRA	<ul><li>Adequacy</li><li>Hazards and risks not identified</li></ul>
Designer report including H&S information	<ul><li>Adequacy</li><li>Hazards and risks not identified</li></ul>
Stage 4: Tender documentation and procurement	
Contractor H&S specification	<ul><li>Additional client H&S requirements</li><li>Hazards and risks not identified</li><li>Additional H&S information requested</li></ul>
Prepare H&S documentation for submission to local authorities	<ul><li>Adequacy</li><li>Approval of first submission</li></ul>
Construction work permit application	<ul><li>Completeness</li><li>Approval of first submission</li></ul>
Facilitation of financial provision for H&S	<ul><li>Adequacy</li><li>H&S items not identified</li></ul>
Site layout	Adequacy
Assessment of the contractor's financial provision for H&S	Effectiveness of the process
Assessment of the contractor's H&S competencies and resources	Adequacy
Stage 5: Construction documentation and management	
Assessment, discussion, negotiation and approval of the contractor(s) H&S plan	<ul><li>Adequacy</li><li>Approval of first version</li></ul>
H&S component of the site handover brief	<ul><li>Adequacy</li><li>Unexpected H&S issues</li></ul>

$\rightarrow$

Criterion/Stage	Form of assessment
Temporary works design	■ Process ■ No failures
Temporary works execution	■ Conformance to design ■ No failures
Welfare facilities	■ Compliance with the Construction Regulations, Facilities Regulations, and H&S plan ■ Adequacy ■ Maintenance
Revised H&S risk profile, specification and budget when scope of work changes	■ Adequacy ■ Lack of incidents or accidents
Implementation of H&S plan	■ Hazards and risks not identified ■ Compliance ■ Accidents, injuries, and incidents
Design HIRA	Design-originated risks encountered on site
Stage 6: Project closeout	
H&S file	■ Process ■ Contributors ■ Adequacy
Implementation of the H&S plan	Compliance
H&S aspects related to operations and maintenance	Adequacy

3.6 Challenges and risks faced and resolution thereof

Challenges and risks faced could include the respective criteria per stage relative to which unexpected challenges or risks were encountered, or could include unique challenges or risks in general.

3.7 Lessons learned

Lessons learned could include the respective criteria per stage relative to which performance was inadequate, problems were encountered or novel responses or solutions were implemented.

3.8 Recommendations for future projects

Recommendations for future projects could be presented per stakeholder per stage.

3.9 Conclusions

Conclusions should flow from the actual assessment, challenges and risks faced, and resolution thereof, lessons learned and recommendations for future projects.

4. Review Questions

- What are the benefits of a construction H&S closeout report?
- Who should compile the construction H&S closeout report, and why?
- What is the relevance of the respective leading and trailing H&S performance measures and the assessment of the respective H&S criteria per stage?

5. Review Exercise

Compile a construction H&S closeout report, or lead a focus group assembled to compile a construction H&S closeout report.

9 781485 132974